GENES

'It's all in the genes.' Is this true? And, if so, *what's* all in the genes? *Genes: A Philosophical Inquiry* is a crystal clear and highly informative guide to a debate none of us can afford to ignore.

Beginning with a much-needed overview of the relationship between science and technology, Gordon Graham lucidly explains and assesses the most important and controversial aspects of the genes debate: Darwinian theory and its critics, the idea of the 'selfish' gene, evolutionary psychology, memes, and genetic screening and modification, including the risks of cloning and 'designer' babies.

He also considers areas often left out of the debate over genes, such as the environmental risks of genetic engineering and how we should think about genes in the wider context of debates over science, knowledge and religion. Gordon Graham asks whether modern genetics might be tempting us to 'play God', and whether the risks of a brave new genetic world outweigh the potential benefits.

Essential reading for anyone interested in science, technology and philosophy, *Genes: A Philosophical Inquiry* is ideal for those wanting to find out more about the implications of genetics and the future of biotechnology.

Gordon Graham is Regius Professor of Moral Philosophy at the University of Aberdeen. He is the author of *The Internet: A Philosophical Inquiry* (Routledge, 1999), *Philosophy of the Arts*, second edition (Routledge, 2000) and *Evil and Christian Ethics* (2001).

'A provocative and always interesting addition to the literature on genetics.'

Arlene Judith Klotzko, *Writer in Residence, Science Museum*

'Clear and level-headed . . . he has a lot of interesting and provocative things to say about both genetic science and genetic engineering. In particular, he has provided as clear and compelling an account of what is meant by "playing God", and what is wrong with playing God, as I have seen.'

Stephen Clark, *University of Liverpool*

'. . . an excellent, readable and lively "take" from the viewpoint of a moral philosopher . . . I think Graham's book will greatly help to clarify the minds of many people who are confused by the media's approach, especially television, that often presents a kaleidoscope of different opinions.'

Dr David Galton, author of *In Our Own Image*

GENES

A philosophical inquiry

Gordon Graham

London and New York

First published 2002
by Routledge
11 New Fetter Lane, London EC4P 4EE

Simultaneously published in the USA and Canada
by Routledge
29 West 35th Street, New York, NY 10001

Routledge is an imprint of the Taylor & Francis Group

Typeset in Meta by M Rules
Printed and bound in Great Britain by
Biddles Ltd, Guildford and King's Lynn

British Library Cataloguing in Publication Data
A catalogue record for this book is available from the British Library

Library of Congress Cataloging in Publication Data
A catalog record for this book has been requested

ISBN 0-415-25257-1 (hbk)
ISBN 0-415-25258-X (pbk)

FOR ALISON

CONTENTS

PREFACE

'It's all in the genes.' One hundred years ago, this little sentence would have been meaningless. Today it has almost the character of a commonplace. And that is why it is important to ask: is it true? As soon as we begin to wonder whether it is true, it becomes clear that some clarification is needed. *What's* all in the genes? And what does it mean to say that it's *all* in the genes? As I hope this short book will show, seeking this clarification quickly takes us into issues that are both important and topical. These include not just the familiar debate between evolutionary biology and its critics, the significance of the genome project, or the real prospects for genetic engineering – GMOs, cloning and the like – but also the much wider questions of science, enlightenment and religion, their nature and their place in society. In short, my purpose is not just to ask whether the sentence is true, but to investigate what its truth (or falsity) means for us as human beings.

The perspective I bring to these questions is that of Western philosophy. This enables me, I hope, to stand above the fray, at least to some degree, and especially the fray between science and religion. If it helps to state my attitude from the outset let me say that I am equally unsympathetic to creationism and to any rabidly anti-religious scientism. But whatever my personal sympathies, my chief purpose is to reason with enough detachment to contribute something to understanding both the rivalry and the complementarity that the less partisan will find between the two.

No less important is my hope of doing so in a way that makes these important matters accessible to serious readers who can claim no special expertise. I have unqualified admiration for professionally conducted science and philosophy. But they acquire their greatest significance when they spill out of the laboratory and the lecture room. Accordingly, the ambitious task I have set myself is that of writing about some complex issues in a style that presupposes little by way of a previous acquaintance with them, and yet does nothing to simplify them either. And at the same time, I aim to bring something fresh to their examination.

There are a number of considerable difficulties facing such a task. To begin with, there is the very limited nature of my own scientific knowledge. I am not familiar with work at the 'cutting edge' of genetics, and probably would not understand it if I were. This means that there is the constant risk of misrepresenting and misjudging the present state and future significance of scientific inquiry. In my view this is a risk we simply have to run. The condition of relative scientific ignorance is one that I share with almost everyone, and if we were all to be silenced by it, we could say and think nothing about some of the most important topics of our time.

Happily, non-scientists need not remain in total ignorance. There are a number of scientists who have written on these topics for a more general audience – Richard Dawkins, E. O. Wilson, Stephen Pinker, Michael Behe, Mae-Wan Ho, Kenneth E. Miller, Michael Ruse, Matt Ridley are among the best known – and there is significant scope for a philosophical engagement with what they have written. Sometimes professionals criticize books in 'popular' science, claiming that popularization means simplification. No doubt it often does. But the fact remains that not only do we need people like Dawkins, Wilson and Miller if there is not to be an unbridgeable gap between science and society, it is also true that it is their views that have fuelled contemporary public debate, rather than the theories and discoveries formulated in more arcane scientific journals.

Another major difficulty is this. The literature relating to the topics of even one of the chapters that follows is huge, and to attempt to cover the topics of all four, especially in a short book, is to run the risk of omitting some central issues. In general I have tried not to do this, and have signalled the places where it has proved necessary. One of my aims, and an important one for the publisher, is to produce a book that is not dauntingly long. Inevitably this means ignoring some important questions. But the alternative is to write at a length few people will have the inclination to read.

The other side of the same difficulty is that of simply repeating what others have already said elsewhere. This is certainly a possibility, though not necessarily a defect. My attempt at originality does not lie primarily in the formulation of a radically novel approach to all the issues I address (though I hope there are a few places in which I succeed in doing this), but rather in the assembly of ideas and arguments (many of which may be familiar to those who already know the literature) in a fresh and illuminating way. This is not a book on science and religion, of which there are many, or a book on the ethics of genetic engineering, of which there are also many. Nor is it a book about the social role of technology, of which there are somewhat fewer. The aim here is to combine something of all three so as to reveal the connections between the immensely interesting and important debates that fall under these general headings. Such is my ambition, at any rate. Whether I have succeeded is for the reader to judge, of course.

I am grateful to Tony Bruce of Routledge who suggested that I write this book, and who, as on previous occasions, brought to the process of its completion his heartening enthusiasm as well as his expertise. His invitation has made me acquaint myself with a wonderfully interesting literature, and the comments of the readers he recruited have greatly improved the final text. I also owe a considerable debt to my colleague Ian Booth, Professor of Cell Biology at the University of Aberdeen, for his help and kindness in ensuring that I have avoided any gross errors with respect to scientific fact. It may well be that I have not learned sufficiently from him. Professor Austin Smith of the University of Edinburgh Centre for Genome Research kindly checked the section on stem cell production and use, and the section on genetic trespassing owes a good deal to an unpublished paper by Professor Richard Sherlock of Utah State University which I had the good fortune to hear him read at a conference. My thanks to Jon Cameron who assisted me with bibliographical research and preparation of the typescript.

King's College, Aberdeen
December 2001

1

SCIENCE AND THE SELF-IMAGE OF THE AGE

Icon and understanding

Contemporary society differs from previous periods in part by the extent to which the printed word has been displaced by the visual image. People still read books in very large numbers, of course – witness the astonishing sales of *Harry Potter* – but the popularity of reading is recreational rather than information gathering. This is primarily a result of the universal presence of television, the cinema, and most recently the Internet. It is also worth noting that high colour printing is easy and cheap nowadays, with the result that very good quality pictures are available in thousands of popular magazines. And thanks to billboards, they are a constant presence in the urban landscape as well. All this serves as a powerful stimulus to the role of the visual in the formation of opinion. 'Seeing is believing' we say; small wonder, then, that the power of the picture is so significant, and can so easily put to one side the impact of critical reasoning conducted in terms of logically structured text and the careful analysis of statistics.

For example. In the 1980s, when concern with animal welfare and animal rights rose high in public consciousness, and with it the cause of vegetarianism, one particular film documentary attracted popular acclaim on both sides of the Atlantic. A striking feature of this film lay in its repeated picture sequences of huge amounts of blood and gore, literally pouring from the slaughterhouses. The resultant image was plain and powerful; these are places of blood in copious, seemingly endless, quantities. But consider what this strictly implies. If our concern is with the pain and suffering of animals – their welfare in short – then, since this was the blood of *dead* animals, it was not evidence of suffering at all. However unpleasant to the squeamish, the removal of skin and internal organs from a corpse cannot be objectionable on the grounds that it causes suffering, since it obviously does not. A dead animal feels no pain. If, on the other hand, the objection to slaughterhouses is couched in terms of the rights of animals not to be killed, then once again the images of blood are irrelevant. Let them be killed as humanely as you like, and let the gore be kept to a minimum, a wrong is still being done to them. To the animal *rights* activist, the manner in which cattle, sheep and pigs are killed is irrelevant. To the animal *welfarist*, the issue is over once the animal is dead.

Either way, the torrential flows of blood the film pictured are not of moral relevance. It is not logical connection but the association of ideas that is at work. But precisely because of this association, the image, as its makers knew, had a decided power to influence opinion in favour of vegetarianism and against the meat trade. In short, *whatever* the truth about animal rights, animal welfare and what goes on in slaughterhouses, the image had a power of its own, and could remain in the minds of those who beheld it more easily and much longer than could any verbal *arguments* that might be mounted.

This sort of example could be repeated innumerably many times, and in quite different contexts. It has been claimed, plausibly, that Hung Cong ('Nick') Ut's one, subsequently famous, photograph 'Accidental Napalm Attack' (1972) altered public opinion about the Vietnam War far more than the thousands of column inches (not to mention books and pamphlets) that preceded it. This photograph showed a naked girl in flames running down a road, apparently *away from* American soldiers and American napalm. The impact of its publication around the world – against American action in Vietnam – was immense. So too with television pictures of civil rights marchers in the Southern states, which, it is widely acknowledged, influenced opinion against segregation more quickly and more easily than almost a hundred years of speeches and pamphleteering had done.

As these last two examples imply (I hope), the impact of the visual may well work to the good. But it is the more general proposition about the power of pictures, for good *or* ill, that I want to underline. This power is extremely important, partly because such influences are prone to generating simplified and sometimes simply false beliefs. Pictures can be highly misleading, and not infrequently a single image influences public opinion only by means of something very like the logically invalid procedure of generalizing from a single instance. However, *this* tendency on the part of pictures is too widely known and appreciated for its observation to count as novel. A more interesting observation is that the potential power of single pictures is just one identifiable aspect of a deeper and more important phenomenon. The power of the visual results partly from the fact that a very great deal of public opinion is structured around and influenced by *abstract* images – images that are complex and generalized – and built up out of both words and pictures. What is specially worth noting is that the sources of these words and pictures are to be found as much in the world of fiction as of fact – story books as well as history books, films as well as documentaries, even advertisements. But whatever their source, they have the power to influence thought and opinion only in so far as they have come to have some common currency. And many of them have.

To take just one instance, popular attitudes to the medical

profession are formed not only by what people know from personal experience or learn from the newspapers (still less from the reports of government sponsored inquiries) but also by dramatic fictional representations – the *Doctor Finlay* novels of A. J. Cronin, for instance (subsequently televised, twice), or the hugely successful American television series *ER*. The same phenomenon can be instanced again and again (with national and historic variations, of course) – the soldier, the statesman, the explorer, etc. and the scientist.

If this is true – that in each age there are stocks of images both created and sustained by an extensive cultural mix of pictorial images, familiar stories both historical and fictional, and artistic representations with greater formative power than academic inquiry, government investigation and public debate – then it is also true that this mixture includes *icons* – which is to say specially venerated and/or specially powerful images. Both images that enjoy a wide currency and those that are iconographic frequently come together to provide society with a *self-image*. That is to say, cultural products of this kind both reflect and re-inforce the ways in which people are inclined to represent themselves to themselves. Such self-images may overlap; what it is to be a contemporary West European usually overlaps with, or at least is incorporated within, the larger image of what it is to be modern minded, to be educated, tolerant and reasonable. Furthermore, since we tend to define ourselves at least in part by how we perceive others to be – the 'others' being distant in either time or space – the result is that the self-image of the 'modern' West is opposed, in common consciousness, to that, for instance, of Eastern Europe as it emerges from the shadow of communism, or of Africa still trapped in a kind of primitivism. Such a contrast, of course, implies a very great deal about the desirability of scientifically based beliefs and democratic pluralism, and of putting an end to tribalism and superstition.

The 'self-image of the age' is a memorable phrase coined by the philosopher Alasdair MacIntyre (MacIntyre 1971), one that accurately encapsulates the concept I am here invoking. Yet the critical mind may reasonably wonder whether it is not too grand a conception, one which at a minimum invites overgeneralization and at a maximum encourages the manufacture of fanciful myths. There is a real basis for this doubt, certainly. People do tend to generalize and to manufacture myths. But this fact about human nature should not divert us from acknowledging that there are indeed powerful 'self-images of the age'. Contemporary Western culture (like all cultures, no doubt) has a self-understanding. It is a self-understanding that can be detected most clearly, perhaps, in commercial advertisements and political pronouncements. But these are merely the outward manifestations of deeply held assumptions that people commonly make, and upon which such advertisements and pronouncements trade.

In fact, the general characteristics of the self-image of the age in contemporary Western culture are not so difficult to discern. Broadly, it has five recurrent features. When we contrast ourselves with ages that have gone before, often of course in considerable historical ignorance, we are inclined to stress our scientific mindedness, our greater technical advancement, the more liberal and democratic nature of our politics, the greater degree of toleration in our moral beliefs, and a freedom from the irrational superstitions and magical cults that marked former periods. Accordingly we look for and commend advances in scientific understanding, technological innovation, democratic procedures, moral pluralism and intellectual enlightenment. All these things are regarded as marks of progress which the modern world both enjoys and enjoins. And this is true even of those who cast doubt upon the universalistic aspirations of what is often called 'the Enlightenment Project' and preach the doctrines of something called 'post-modernism'. They, too, are convinced of their belonging to a more pluralistic and tolerant age.

In short, the contemporary self-image of the age in the Western world, especially among the educated (some would say 'chattering') classes has a sort of composite coherence. There are moments of doubt, it is true. The history of the twentieth century, with its wars, concentration camps and Gulags, does not seem specially suited to sustaining the idea that morally speaking Europe is better than it was. Is there such a radical gulf between 1930s Germany, a modern, technologically sophisticated and scientifically advanced society, and 1990s Rwanda, ill-educated and trapped in poverty and underdevelopment? Both, after all, were scenes of unspeakable carnage allied to racialist hatreds. Even when it comes to the demise of superstition the picture is not altogether clear. Such practices as astrology persist in the West, and are given surprising degrees of credence even in powerful circles. It was said, truly or not, that President Reagan of America, no less than Mrs Gandhi of India, had recourse to clairvoyants and soothsayers. The evidence of religious belief (a subject to be returned to), which is not infrequently regarded as another differentiating marker, is also mixed. If current statistics on church going are to be believed, religious adherence is rapidly declining in Western Europe but still a powerful force in Africa, Asia and Latin America. This is just what the self-image of the age would incline us to expect. Contrary to that image, however, religion is also very much alive in the United States, the most technically advanced nation of all.

There are, then, some doubts about the modernity and enlightenment of the West, but whatever these doubts, which very many people hold I think, in this respect our world is different from those that preceded it, and correspondingly different from many that prevail in other parts of the globe – the Islamic Arab world (with its floggings and amputations), sub-Saharan Africa (with its violent tribalism), and Hindu India (with its castes and sadhus) being particular points of contrast. In short, scientific understanding, technological innovation, democratic procedures, moral pluralism and intellectual enlightenment make up our society's self-image.

The fractured image: Einstein *vs* Frankenstein

However: if science and technology are indeed a central part of this image, it is a decidedly fractured one. On the one hand, scientific technology is widely believed to have brought us the prospect of dispelling human ignorance and advancing human prosperity. It is in this sense that we speak, in commendatory tones, of the superiority of taking 'a scientific approach' to questions of both theory and practice. On the other, this very same scientific technology can seem to threaten death, destruction or possibly something (quite literally) unimaginably bad. These are the resonances that can often be heard around the terms 'nuclear technology' or 'genetic technology' for example, both of which are as frequently regarded with anxiety as with admiration. So, while people trying to cope with famine, epidemics or ignorance readily turn to scientifically based agriculture, biomedical research and hi-tech educational methods, in almost equal measure they fear the polluting effects of pesticides and herbicides, the unnatural monsters that biotechnology might create, and the corrupting impact of television and the Internet.

This fractured image is partly captured by the contrasting icons of Einstein and Frankenstein, and the attitudes to science and technology that they both express and reflect. Though Einstein's reputation has come in for criticism of late – he seems to have treated his wife quite badly and encouraged the development of the atomic bomb – even so, the familiar white-haired picture of the most famous of scientists almost defines the benevolent and the wise. The popular use and reception of the image is that of a decidedly kindly man, who despite his avuncular appearance, is the very embodiment of the ability of the scientific mind to break through to astonishing new heights of explaining and understanding the physical world. In short, this is the face of a good genius, and it explains the way in which the boy (or girl) genius is approvingly called 'an Einstein'. By contrast, though there are some endearingly comic resonances (in *The Munsters*, for example), in general the fictional image of Frankenstein, and the ultimately uncontrollable monster he created (replicated in a different technology by the computer HAL in Stanley Kubrick's film *2001: A Space Odyssey*) is the archetypal image of the *evil* genius. Frankenstein was the invention of Mary Shelley. Jon Turney has shown how well versed Shelley was in the science of her day, and how this played a crucial part in generating a mythology of huge cultural importance, a mythology constructed in large part by fiction but no less influential in the reception of fact (see Turney 1998 *Frankenstein's Footsteps*). Indeed, the name of Frankenstein is regularly deployed to express, and to invoke perhaps, the recurrent fear that science may overreach itself – that genetic technologists are creating 'Frankenstein foods', even that they are 'playing God' or that atomic physicists, in an excess of enthusiasm for science and all its works, have brought into existence a force – nuclear power – that threatens mass destruction, in its civil no less than its military manifestation. Indeed, so powerful has this second fear been that the world has largely turned its back on a source of energy more environmentally 'clean' than any other, and whose actual (as opposed to potential) harm is very much less than the sickness, death and damage caused by the use of fossil fuels.

That the popular understanding of science and technology is Janus-faced in the way just described can hardly be doubted. The icons of Einstein and Frankenstein are equally prominent in our consciousness, and pull us in opposing directions. Just why this should be so is an intriguing and difficult question. It is one for history and sociology, perhaps for psychology, to answer. By contrast, the philosophical question with which I am concerned here has to do not with the rise of these alternative images, or with their persistence, but with their *justification*. What truth and substance, if any, is to be found in them? Can either image be given a convincing philosophical underpinning that will withstand serious criticism? Which ought we to adhere to? Or should we, rather, abandon both in favour of some more intelligible and intellectually defensible alternative?

These are crucial questions for contemporary self-understanding. My contention in this short book is that serious thought must lead us to dwell at length on this particular aspect of the self-image of the age, and modify our wider understanding of science and technology.

Science

There is a successful and enduring textbook in the philosophy of science (by A. F. Chalmers) with the arresting title *What is This Thing Called Science?*. This is not only a good title; it is a good question. In setting out to answer it, it is worth observing that the terms 'science' (as the name of a particular discipline) and 'scientist' are of relatively recent date – post-1830, roughly. Indeed, Chalmers's question, arguably, could only be the title of a book in the English language since it is only in English that 'science' is regularly used in the singular. Other European languages have generally referred to 'the sciences'.

This is not merely a linguistic oddity, but a difference of some consequence. The Latin origins of the word 'science' imply nothing more ambitious than 'knowing' and 'knowledge'. Accordingly, until modern times 'a science' meant simply a body of knowledge and, more sophisticatedly, the distinctive methods by which it was acquired. The eighteenth century could speak happily about 'the moral sciences', terminology that is preserved to a limited extent in some older universities. It is an expression which strikes us now as archaic. But in its day it was novel, and carried an implication that over time gathered strength, to the point where science in the singular came to mean a specially good sort of knowledge, and a 'scientist' came to be someone who spoke with a certain sort of authority. This authority is implicitly invoked, in fact, when the media begin a report with the words: 'Scientists have discovered. . .'. The (usually unspoken) assumption is that what follows is not just opinion, but expertise and knowledge. And the fact that no such reports ever begin 'Philosophers (or theologians) have discovered. . .' reveals the further assumption that there is just one form of inquiry that generates real knowledge, only one really rational kind of understanding.

We can better appreciate this important change if we recall that the subject known today as 'physics' – to many minds still the prime example of science – was once known as 'natural philosophy' (as, once again, it still is in a few older universities). By contrast, nowadays anything called 'philosophy' is viewed quite differently from anything called 'science'. This is illustrated by the fact that in contemporary English it is permissible (though in my view erroneous) to speak of 'my philosophy', while 'my science' is an expression without any obvious use. Now just as 'natural philosophy' gave way to 'natural science', so the eighteenth century tried to replace 'moral philosophy' with 'moral science', and thereby introduce (as David Hume records) the methods of Bacon and Newton to the study of moral subjects. In this case, however, the change did not stick. 'Moral science' has a decidedly quaint ring to it. The name 'moral philosophy' returned, though it has come to be overshadowed of late by references to 'ethics'.

These brief observations about the history of linguistic usage reveal a significant development in the history of ideas. While previously *all* forms of inquiry were regarded as 'sciences', now only *some* are. The change is most marked in the case of theology. Where once it was possible for people to claim that theology was 'the queen of the sciences', now, far from being their queen, theology has been thrown out of the company of the sciences altogether, and the word 'theological' is used (outside the discipline itself) to mean the arcane and obscurantist. In modern parlance 'theological' just means the realms of the dogmatic and unreasoning. This supposed contrast is most evident in the high-profile clash between evolutionists and creationists (to be considered at some length in the next chapter), but its influence can be found at work far beyond the confines of that particular debate.

This narrowing and confinement is not just a matter of classification. 'Scientific' is an accolade. It has normative as well as classificatory force, and it is because it has this normative dimension that the question 'Are the social sciences, sciences?', for example, can be raised and discussed with some intensity. Creditworthiness turns on the answer we give. If the social sciences are not sciences, they cannot be held in the same esteem. In contradistinction the question 'Are the natural sciences, sciences?' cannot be asked to any point or purpose; physics, chemistry and biology are just assumed to be sciences *par excellence*. Their status is not in doubt because they are themselves the benchmarks.

Our common understanding of the nature of science has a further dimension to it. If philosophy is nowadays regarded as 'personal' in a way in which science is not, and theology has been banished to the realms of dogma, something more needs to be said about mathematics, which does not easily fit into these contrasting categories. Unlike philosophy and theology, by and large mathematics has retained its intellectual creditworthiness, but is not as clearly regarded as a branch of science. Why not? The difference, I think, is that whereas pure mathematics is widely regarded as 'useless' (if nonetheless admirable), 'science' is more favourably viewed as 'relevant'. But wherein does the 'relevance' of science lie, exactly? A common answer is not far to seek. Science is 'relevant' because as it progressively develops, it gives rise to technologies that increase economic prosperity and advance human well-being. Pride of place in this respect generally goes to medical science – 'the greatest benefit of mankind'. This goes some way to explaining the colossal sums of money that the modern world spends on medical research, sums that far exceed those spent on other subjects. It also explains why politicians and the public can be induced to worry about the contraction of scientific research, when they could never be persuaded to worry about the contraction of philosophical or theological inquiry. The voguish term 'knowledge economy' tacitly supposes that the knowledge in question is a knowledge of genetics or computer science, not a knowledge of classical languages, Hindu mysticism or the metaphysics of Thomas Aquinas.

These are broad brush remarks, it must be admitted. I have said nothing about history, economics or psychology, none of which fits so very obviously into this general story and about which there are interesting things to be said. But I believe that the picture of 'science' I have been painting, and its relation to other intellectual pursuits, does capture a common way of thinking about its nature and value. Yet it is not difficult to see that there is a good deal of confusion and error in it. Among the greatest of those discoveries attributable to science (which is to say the natural sciences) are several that have no practical relevance at all. The marvellous and illuminating investigation of theoretical physics and cosmology that began with Sir Isaac Newton is a plain case in point. Why do apples fall to the ground? Why does water invariably flow downhill? Why does the moon go round the earth? Why does the earth go round the sun? Why is the sky dark at night? These seemingly simple questions have led to the most extraordinary hypotheses – that there can be action at a distance (i.e. one thing can act on another without ever touching it); that the universe is both infinitely large *and* expanding; that the cosmos began with a 'big bang' that brought time itself into existence.

Fascinating though these hypotheses may be, and they certainly constitute excursions of the human mind that reveal it to have truly astonishing powers of understanding, they are of no practical relevance whatever. People knew that water only flowed downhill, and harnessed this fact to their advantage in aqueducts and watermills, long before they understood anything about gravity. Likewise, there is nothing we can do that we could not do before in virtue of the fact that we now know the universe to be infinitely large and at the same time expanding. What could there be to do? How could our knowledge of the expanding universe be put to good use? The question is rhetorical, of course. It couldn't.

This is not just true of cosmology, however; it is true of biology also. Charles Darwin, rightly, is regarded as having initiated a huge shift in our understanding of the natural world. Thanks to his careful and protracted inquiry, we now know, *pace* the creationists, that the species of plants and animals that populate the world have evolved over time. There is still contention over the mechanisms Darwin postulated – random mutation and natural selection (both topics to be considered further) – but one thing seems to me certain; to know that species have evolved over periods of time longer than any previously conceived, is not to be in a position to do anything that we were not in a position to do before. Certainly, the advent of genetics, originating with the obscure monk Georg Mendel, combined with evolutionary theory, is widely thought to have profound and widespread practical implications. It is a subject with which we will be concerned at length in later chapters. But whatever the truth about the significance of genetic biology, just to *know* that there has been a long evolutionary process of which human beings are themselves a part is a massive contribution to, not to say revolution in, human understanding. Even so, it is an advance that leaves us no better off, practically speaking, than we were before.

Even medical science is not without its 'useless' episodes. William Harvey's discovery of the circulation of the blood, in and of itself, contributed nothing to therapeutic advances. Indeed, it is a striking if little remarked fact that while over the course of several centuries (culminating in the pre-eminence of the University of Edinburgh in the eighteenth) anatomical investigations made huge strides, medical treatment remained largely ignorant and primitive for another hundred years or more. The most innovative and original anatomists did almost nothing for the advancement of human health. Harvey is one of the great names in medical science; so too is Alexander Fleming, the discoverer of penicillin. But there is this notable difference between them; Fleming's discovery had dramatic implications for the treatment of disease; Harvey's did not; and a proper understanding of the nature of science must accommodate both sorts of discovery.

In short, whatever the merits upon which the superior nature of 'science' may be based, practical benefit cannot always be one of them.

Technology

Newton, Harvey and Darwin are stars in the scientific firmament, in the whole of intellectual history indeed, and it is largely from the existence of such names that science derives its contemporary *kudos*. But this very fact demonstrates that the connection between science and technology cannot quite be as it is commonly supposed to be. It is certainly true that *some* scientific investigations have resulted in enhanced technological ability – Fleming's is a specially striking instance – but this is not true of all, or even perhaps of many. Conversely, a great number of the advances generally heralded as 'technological' have been made largely in ignorance of the scientific explanations that underlie them. This is an equally important point to underline. For example, the invention of the printing press, the spinning jenny, the steam engine, the telephone, the motor car, the combine harvester, the aeroplane, hire purchase and the credit card (as opposed to the swipe card), have contributed hugely to economic prosperity. Yet their invention owed relatively little to the investigations of science as we normally understand it. All of them are rightly referred to as examples of 'modern technology'. But if 'technology' does not necessarily have its origins in 'science', where does its distinctiveness lie?

A well-known advertising slogan (from the Italian manufacturer Zanussi) neatly captures a widespread idea; the technology of the washing machine and the fridge, it tells us, is 'the appliance of science'. Yet if the facts I have just been adducing in this section and the last are correct, this simply is not true. Many major scientific advances of the modern period have thrown up no new technologies, and the technological advances that have been made in the same period owe relatively little to scientific investigation. All these devices harness physical forces which natural science has been highly successful in explaining, certainly, but they are the products of inventors, not of scientists. Atoms, molecules, radio waves and the rest have to behave in certain ways and with law-like regularity if petrol is to be converted into power, if satellite communication is to be possible, and if refrigerators are to keep things cool.

But if technology is not in any obvious sense 'the appliance of science' what does mark it out? Answering this question requires us to explore briefly the differences between technologies and techniques. Consider this relatively mundane example – washing clothes. Since time beyond recording, people have washed their clothes. We can still see the techniques they have used in the poorer societies of the contemporary world – beating fabrics against rocks on the riverbank. What makes this, compared with the modern washing machine, a technique rather than a technology? What makes the vacuum cleaner an advance on the sweeping brush? The answer is relatively simple, I think. Technology is the use of machines to perform practical tasks. A technique is a way in which human beings act directly to bring about some desired result; technology is the use of machinery to achieve the same or similar ends more effectively. Thus sewing and knitting are techniques to make clothing; sewing machines and knitting machines are technologies that make clothing more efficiently.

It might be replied that this cannot be the whole answer. Why are sewing and knitting needles not technological devices (of a simple sort) also? And why is a vacuum cleaner to be graced with the title 'technology' when a brush is not? Again, as it seems to me, there is an obvious answer: machines employ mechanisms, interacting parts that are related functionally. This is what allows us to speak of technology as both 'operating' and 'being operated'. The use of a knife to cut is a technique; the knife in cutting neither operates, nor is operated; it is simply used. If I use it correctly, it cannot fail to do what it is supposed to. We may describe the use of a bacon slicer, on the other hand, as the introduction of technology to the same procedure because, unlike the knife, despite my best efforts it may fail to operate properly; the functional relationship of parts within it may go awry independently of my using it.

This criterion of distinction, though simple, proves surprisingly effective. We have only to think of anything that intuitively one might want to call 'a technology' and it will be found, I think, to meet this account of what it is for something to be a machine. Of course, this account of the matter is not meant to introduce or justify stipulative definition. We can *call* a knife or a brush simple technological devices if we wish. The point of distinguishing techniques from technologies in this very straightforward way lies in its ability to clarify our understanding of science and technology.

A serious doubt about this characterization of technology might be thought to arise when we consider what is generally known as hi-tech, and in particular biotechnology and information technology. Surely the idea that technology is a matter of manufactured devices with mechanisms takes its cue from the printing press and the motor car, both of which obviously have moving parts, and not from silicon chips or genetically modified tomatoes, which equally obviously do not? Partly this doubt rests on a mistaken assumption about what 'mechanism' means. Certainly, the term comes from 'mechanics', the construction of physical machines with cogs, wheels, pulleys, levers and the like, and everyday language may still carry this connotation. But the concept of a mechanism in itself need not imply that its component parts are physical in the way in which the spring, cogs and balance of a clock are. So far as the present analysis is concerned, it is as proper to speak of biological as of physical mechanisms. The power of the organic no less than the inorganic can be harnessed to practical purposes. Holmes Rolston has observed that 'the transition from muscle and blood, whether of human or of horses, to engines and gears shifts by many orders of magnitude the capacity of human beings to transform their world. Even more recently the capacity to produce has been augmented by the capacity for information transfer' (Rolston 1998: 5) and the same point can be made about the advent of organic mechanisms – what we now refer to as biotechnology.

Even so, there is plainly something to the contention that hi-tech stands in a special relation to the science that underlies it; that what we witness in it is a 'step-change' compared with older technologies. The difference is this perhaps; the development of hi-tech is impossible without a sophisticated understanding of the science that underlies it because it relies upon the extension of that science. Modern biotechnology is just of this kind; advances in engineering just are extensions of the science that explains how they work. Yet even here, the failure to draw and preserve a distinction between science and technology can result in a measure of confusion. In public discussion of the importance of science, and public concern about new technologies, very often those who are anxious about technological developments feel bound to cast doubt on scientific explanation, and conversely, those who welcome such developments pride themselves on being scientifically minded. Consequently, it is of the greatest moment to see that even in the world of hi-tech, science and technology are not simply to be lumped together as one composite whole. Indeed it is only if we understand the relation between the two to be far more nuanced than this, that we will be in a position to see just what a genuine admiration for scientific understanding implies when we turn our attention to what appear to be world-altering technologies.

Having separated science and technology in this way, the next question to be raised is this: What is it that has given 'science' the status it has, its pre-eminence among all the forms of inquiry in which human beings engage? The short answer is its power, and contrary to the distinction I have just been at pains to underline, this is generally regarded as a twofold power – the power to explain *and* the power to engineer. Science, it is widely believed, can explain phenomena in ways in which other kinds of inquiry cannot. 'What is the scientific explanation of X?' is generally understood to mean the same as 'What is the real (or true) explanation of X?'. It is for this reason that science is believed to relegate theological explanations to the realms of magic and superstition, and to render philosophical explanations as 'airy fairy' if not positively mystical. In short 'the scientific' and 'the rational' come to the same thing. In a similar fashion 'scientific' solutions to practical problems are thought to be real solutions. Modern medicine replaces spells, incantations and 'old wives' tales' (as well as prayer and faith healing) by being both rational and effective. More importantly yet, it is effective precisely because it is rational.

Anti-science

This additional contention captures, I think, the essence of the modern understanding of science and technology. Scientific explanations are rational explanations, and modern technology is effective because it is based upon rational explanation. It is a contention that has had a powerful influence on the shape of contemporary culture. It is also a contention that it is part of the purpose of this book to examine critically. Examining it critically, however, is not the same as discounting it or condemning it. It is no part of my intention to espouse a 'post-modernist' view that aims to bestow on 'the poetic' and 'the mythological' equal status with 'the scientific' on the grounds that they are all only culturally relative ways of weaving words anyway. It is true, as Brian Ridley has observed, that 'early scientists like Gilbert and Newton were firmly in the magical realm' and true that we can legitimately ask how far science has left magic behind (Ridley 2001: Ch. 4). But it is necessary to accept something which I think to be obvious – that the natural sciences have made real advances in human understanding, and that in large part they have done so by replacing the magical and the mythological.

Neither do I mean to defend a more subtle 'anti-scientism', or to deride modern technology. There are those who do, of course. Prominent among such people is the American essayist Wendell Berry, whose most recent book – *Life is a Miracle* – aims to assert (among other things) that 'science' falsely dispels the miraculous and ultimately mysterious nature of human life and the world in which it must be led, and thereby impoverishes it. Berry wants to oppose a scientific view of the world, and in so far as he is challenging some of the overweaning ambitions that enthusiasts for 'science' often have, there is merit in his case. My concern, however, is not with 'science' writ large, but with the limited case of genetics, and my question is whether there is reason to think that the genetic revolution in science and technology is all that it is made out to be.

Nor is it my aim to give succour and solace to those (Berry partly in some of his writings) who want us to turn our back on 'modern technology' because of the destructive and corrosive impact they see it as having. The most dramatic instance of such a stance was that of Ted Kaczynski – the so-called Unabomber – who directed his destructive attention both to universities as centres of 'science' and to airlines as the users of advanced 'technology', before he was caught and gaoled. (I discuss this case at slightly greater length in Graham (1997) Ch. 2.) Kaczynski's actions were seriously life threatening and attracted notoriety for this reason, but his objections to modern technology, published as part of a negotiation by the *Washington Post*, are far from being the ravings of a madman. Indeed his position is, if anything, more articulate than that which lies behind the more moderate 'direct action' of those 'green activists' who have set about destroying fields of genetically modified crops. Though their actions may be less dangerous, they share with Kaczynski a less well worked out conviction that modern technology threatens us with very great dangers. In a subsequent chapter I will examine in closer detail the issues surrounding genetic modification, but here I merely record my view that, since biotechnology is a fact that will not go away, the real issue is not how we might forestall or prevent it, but how best we are to understand it.

Berry and Kaczynski, though wholly at variance in their methods, can each be thought to be a proponent of the negative image of the scientist, people who (the first in words, the second in actions) seek to counter the positive image that popular and establishment opinion about science assumes. It is not only in their responses that they differ, however. The focus is different too. Whereas Berry is in large part concerned with the *hubris* of something called scientific *explanation*, Kaczynski was primarily attacking (as he saw it) the bogus credentials of modern technology as a source of *improvement*. In other words, their respective positions cut across the Einstein/Frankenstein contrast I pointed to earlier, and mirror, indeed, contrasting ambitions that the proponents of 'science' on the opposing side tend to have. Berry's doubts (or many of them) are about the ability of modern science to explain; Kaczynski's obsession was focussed on the ambition of modern technology to enhance and improve. These are not the same. It is the concern about technology that most readily allies itself with the image of Frankenstein. From the laboratory of Frankenstein monsters emerge; from the studies of Einstein come only theories. The cost of erroneous theory is merely absurdity; the cost of erroneous application is harm. What is commonly called science, then, whether we view it positively or negatively has two aspects – it aims to explain things, and it aims to engineer things.

Explaining things

In an earlier section I drew attention to the fact that some of the greatest scientists made discoveries of no practical import, but they explained the phenomena with which they were concerned more fully and more adequately than it had proved possible to do before. It is partly this striking ability to explain that began to put natural science ahead of what are often seen as its rivals. When the French scientist Laplace said of God, 'I have no need of that hypothesis' (he meant in his scientific explanations), it was not difficult to hear in his remark the further thought that perhaps no one did. Now no one *did* need it (the God hypothesis), if two propositions hold; first, that natural science can do without the concept of divine agency; and second, that there is nothing natural science cannot explain. People commonly suppose that these two contentions go together, that the irrelevance of theology to science and the adequacy of scientific explanation are sides of the same coin. But logically speaking, it is possible that the first of these propositions (science does not need to import the idea of God into its explanations) holds true, while the second (science can explain everything) is false. If science *cannot* explain everything, whether the things it cannot explain are to be explained by appeal to divine agency or in some other way is a further question. The important point to grasp for present purposes is that it is only if *both* hold true that science can properly be accorded the pre-eminence it has in our culture. It may well be that science has the last word in some things; it is another matter to claim that it has the last word in all.

To examine the issue properly we need to be clear about what exactly 'science' and 'scientific' mean. I have been arguing that the pre-eminence of natural science over other forms of inquiry is both of relatively recent date, and a presumption rather than a conclusion of most thought about the matter. This presumption can be given a slightly more precise form, however. To believe that 'science' is the only form of inquiry and explanation is (generally) to regard the natural sciences as the best model for *all* inquiry. To do so is to subscribe (unknowingly to most of the subscribers, no doubt) to a certain epistemology or theory of knowledge – that the truth is most likely to emerge from empirical investigation, which is to say, investigation into perceptually observable fact, and that only that which can be tested by such means as the telescope and the microscope is ultimately to be relied upon.

Some readers will think this to be a caricature of science.

Probably it is. Empiricism is a doctrine whose weaknesses philosophers have regularly exposed. It seems both unjust and implausible that the practice of science, given its real successes, should be saddled with it. But I believe that the empiricist presumption captures a common conception, albeit one that simplifies and distorts what is actually happening in the real life of physicists, chemists, geologists, zoologists, botanists, microbiologists and the rest. However this may be, the idea that 'science' so conceived holds the key to every mystery is an assumption made by many people. Such people are more likely to be non-scientists than scientists, I am inclined to think; it was the popularizer T. H. Huxley for instance (though he was not only a popularizer), who made the greatest claims for the theory of evolution rather than Darwin himself. But whether made *by* or *for* scientists, we ought to ask whether the assumption is a defensible one. Modern cosmologists and theoretical physicists regularly speak of their ambition to formulate a 'theory of everything' (TOE), something they have not yet done, of course. Upon examination this dramatic expression indicates a more limited ambition than first impressions might suggest. The TOE, it turns out, even if we had it, would leave a very great deal *unexplained*, not least the working of the microbiological world. In fact, a greater degree of ambition is to be found for and among biologists. There, by contrast, it does seem that many see in modern biology, if not a theory of everything, at least a theory that will explain a very large number of hitherto disparate phenomena – physiological characteristics, health, life expectancy, psychological traits, personal aptitudes, social behaviour and even the basis of human nature itself. What determines the distribution of blue and brown eyes? Why are some people susceptible to cancer, or to heart disease? What makes people musical? Why are males more aggressive than females? Wherein lies the difference between human beings and other animals? These are all interesting questions that we can hardly fail to be intrigued by, once they have been raised. But where do the answers lie? Until very recently it has generally been supposed that the explanation of such apparently disparate phenomena must come from different sorts of inquiry – from physiology *or* psychology *or* sociology, for instance. What recent excursions in biology seem to suggest is that the answer to *all* of them lies in just one source – our evolved genetic structure – and that to understand this structure and its evolution is to understand a very great deal about animal life and human existence. The scope of this ambition on the part of a single science – to have uncovered the underlying explanation of things as diverse as eye colour, disease and social behaviour – is reflected in the names of new fields of inquiry that have emerged from it, sociobiology and evolutionary psychology being the best known. Counted among its enthusiasts are some of the best known contemporary science writers – E. O. Wilson (*Sociobiology*), Stephen Pinker (*How the Mind Works*) and above all Richard Dawkins (*The Selfish Gene*, *The Blind Watchmaker*). Dawkins is unequivocal about his position. The subtitle of another of his books – *River out of Eden* – makes the point starkly – 'a Darwinian view of *life*' (my emphasis) – and in the preface he expressly asserts that 'In one way or another, all my books have been devoted to expounding and exploring the almost limitless power of the Darwinian principle' (Dawkins 1996: xii).

Can such a universalistic ambition be justified? What appears to make it plausible are the huge strides that have been made very rapidly by a newcomer to the scientific stage – genetics. In combination, the discovery of the double helix structure of DNA by Francis Crick and J. D. Watson (in 1953) and the completion of the human genome project (in 2000), against the background of Darwinian evolutionary theory, which turned mere naturalists into biological scientists, are widely thought to have constituted a breakthrough in human understanding so great, that we are on the verge of formulating explanations deeper and more wide ranging than has ever been possible before. Are the enthusiasts such as Dawkins right? Suffice it to say at this point that there are dissenting voices, and dissenting voices from within biology itself no less than from anthropology, philosophy, psychology and sociology. The contention, and the objections to it, will form the subject matter of the next chapter.

Engineering things

There is, however, a further and distinct dimension to this issue. Are we also on the verge of *fashioning* a new world and remaking the vegetables, animals and human beings of the future as we will? To some these are exciting possibilities that the new science of genetics has made possible. To others they are more a nightmare than a dream. (*Genetic Engineering: Dreams and Nightmares* is the title of a useful scientific introduction by Enzo Russo and David Cove; *Dream or Nightmare?* is the subtitle of Mae-Wan Ho's campaigning book *Genetic Engineering*.) Both sides often suppose, however, that the success of the science of genetics automatically has implications for engineering, that the two go hand in hand. This simply is not so. Even if evolutionary genetics proves to have explanatory power far beyond anything that preceded it, it does not follow (if what I have been saying about science in general is correct) that we are thereby possessed of an equally powerful technology. At the risk of undue repetition let me say again that many of the most important advances in the sciences have been of no practical significance whatever. It may be, therefore, that *science* is far more powerful than any associated *technology*. In fact this is true, and easily demonstrated, in the case of genetics itself. The most famous names in modern biology are those of Crick and Watson, the two Cambridge scientists who discovered the existence of DNA, and were rightly rewarded with a Nobel Prize for the immensity of their discovery. But by itself, the enormous significance of this discovery did not place any new practical power in our hands. What did so were some technological breakthroughs that came a little later – the ability to use restriction enzymes to cut DNA into gene-sized pieces, the ability to separate different pieces of DNA, and the ability to force one DNA piece into a bacterial cell. Without these techniques, the discovery by Crick and Watson could not have led to the medical and related developments that are now such a prominent topic of public discussion. In other words, in genetics as in the other sciences, there is a distinction to be drawn between the power to explain and the power to engineer. With respect to the latter a different set of questions arises, many of them ethical, and these will form much of the subject matter of Chapter 3.

There is, though, this point of difference to be recorded. In general science and technology should be distinguished in a way that makes it clear that advances in the former are no guarantee of advances in the latter, and, as I have argued, advances in technology may not depend upon scientific knowledge. Edison was able to invent the light-bulb in *relative* ignorance of the physics of electricity, just as Watt invented the steam engine without needing to know Boyle's theory of gases. With respect to modern biotechnology, however, this relative independence of science and technology is much less clear. While it is true that the technology of *in vitro* fertilization ('test-tube babies') has been of very great importance in enabling scientists to modify organisms and tackle hereditary diseases and yet did not itself rely heavily on a knowledge of genetics, no one could have discovered the means to cut up DNA without being deeply versed in the discoveries of Crick and Watson. In short, there is a more intimate connection between science and technology in the case of genetics than in many other cases of technological innovation.

In this respect genetics is not unique. The technology of the

Internet could not have been developed by people who lacked an extensive knowledge of computer science. This is the mark of 'hi-tech' as I earlier characterized it. Whereas with other technologies – the motor car, the telephone and so on – the science that explains them plays no direct part in the engineering that devises them, in hi-tech, information technology and biotechnology being the obvious cases, effective engineering requires extensive scientific knowledge. Unquestionably it is the case that only those who know a lot about the science of genetics are in a position to develop genetic therapies.

The converse does not hold, though – not all experts in genetics are genetic engineers – and this is enough to show that the two are conceptually distinct. Science and technology are different things. Moreover, even in the case of hi-tech, they are to be distinguished to some philosophical point and purpose. Genetic science and genetic engineering present us with different ambitions and different possibilities. Accordingly, the hopes and fears associated with these ambitions and possibilities are different. In particular, the claim of evolutionary biology, enhanced by genetics, to be able to explain with 'almost limitless power' is a different claim from that of the engineers who promise us the power, say, to eradicate hereditary diseases. Consequently, the issues they raise need to be addressed independently, if only for a time, and this is why they form the subject of two different chapters – the next two in fact. Before that, however, we need to say something about the underlying concepts of use and value.

Pure and applied science

There is an important distinction to be drawn among the activities in which human beings engage, between those whose purpose is to sustain life and others whose purpose is to make life worth sustaining. We might express this distinction as one between useful activities and valuable ones. The same distinction can be drawn among artefacts and other possessions – some objects are useful, others are valuable. Of course any given activity, or object, can be *both* useful *and* valuable. Even so it is easy to see that we have to draw some such distinction because, whereas we can always ask of any activity (or object) that is said to be useful – what is it useful *for*? – to ask what something is valuable for, is to reveal a confused way of thinking.

Now some people think that the familiar distinction between pure and applied science is to be explained in these terms – applied science is useful, but pure science is valuable in itself – 'knowledge for knowledge's sake' is the usual slogan repeated in this context. But though not wholly inaccurate, this way of thinking can incline to ways of thinking that are deeply misleading. The first is this. 'Useful' is not a positive term, strictly speaking, but a neutral one. That is to say by correctly declaring something to be useful, we have not thereby given anyone reason to commend it. Landmines are useful for injuring people at random; atomic bombs are useful for destroying major cities at a go. Both these propositions are true; the devices in question do have these uses. But I take it that to admit this usefulness is quite consistent with believing that the world would be much better off without them.

A second point of some consequence relates to the valuable. The slogan 'knowledge for knowledge's sake', like the slogan 'art for art's sake', is a mantra often repeated by the defenders of pure science and the high arts as though its repetition could by itself bestow value on the things in question and thus put an end to the utilitarian carpings of critics. But this simply is not so. There can be trivial knowledge, and what is more, trivial knowledge that people are genuinely interested in. Train spotting is sometimes cited as the classic example. Train spotters (if there are any left) gather considerable quantities of information, as a result of empirical fieldwork, in which, it seems, they have a passionate interest. Pressed to say what use all this information is, they may pronounce their guiding maxim to be 'train spotting for train spotting's sake'. Yet while they are right to resist the utilitarian's incoherent idea that only the useful can have any value, their invocation of this principle does not elevate their activity into the realms of the sciences. In short, something more needs to be said about why cosmology, say, is to be ranked far above trainspotting in terms of its value; the pursuit of knowledge is not enough to distinguish the two.

The upshot of these points is this: it is not enough for evolutionary biologists to tell us that they are adding to our knowledge. The humblest gatherer of the most mundane facts does that. Rather they need to show us that the things they discover add up to theories and hypotheses that have the power to generate explanation and understanding on a far greater scale than hitherto. And of course, most of them do claim this. Indeed, the idea that modern study of the natural world has undergone a qualitative shift in this regard is reflected in the fact that the old terms 'botany' and 'zoology' have widely been replaced with 'plant biology', 'microbiology' and 'life sciences'. This returns us to one of the themes with which we began: does science (narrowly understood) warrant the veneration that the image of Einstein suggests? More especially, does biology deserve this status?

This is a question about the *value* of modern biology. It can be asked irrespective of any interest in its *use*, and will in fact occupy us through the course of the next chapter. However, there is also a question of usefulness. Are the investigations of modern biology useful? To many people, I imagine, the answer is glaringly obvious, as indeed I think it is. Modern biology has enabled us to do, and to consider doing, many things that were hitherto not only impossible but unimaginable. Recalling my point about the neutrality of the term 'useful' and the ever present question 'useful for what?', we cannot simply suppose that its usefulness brings with it a commendation. The image of Frankenstein is not that of the alchemist, the uselessness of whose science is revealed by its repeated inability to turn base metal into gold. No, Frankenstein's science has effects; the question is whether the effects it has are to be valued or feared. It is easy to see how these questions arise with respect to biotechnology and genetic engineering, and these will form the subject matter of Chapter 3.

One way in which people naturally seek to show that the useful is valuable is to show that it is a more efficient or cost effective way of getting what we want. This makes the final touchstone of applied science the satisfaction of human desires – for better food, longer life, improved health and the like. This may sound innocent enough if we stick to these particular desires (though there is more to be said about them also), but in fact it is precisely at the level of human desires that some of the most troubling questions surrounding modern biology arise. This is because human beings can entertain very ambitious desires. In times past such ambitions, like Frankenstein's monster, were confined to the pages of fiction. Or, they made slightly ridiculous appearances in the suggested techniques for better breeding that eugenicists advocated and eccentric millionaires tried to put into practice. But with the advent of genetic technology these ambitions take on a different character and raise prospects which alarm some as much as they enthuse others – 'designer babies', 'playing God' and so on. These familiar phrases, which headline writers find so handy, may or may not point us to real possibilities and dangers. Whether or not they do is partly a question of engineering – can the thing actually be done? – but also a matter of the fundamental values that should confine and structure our genetic aspirations. This will be the topic of the fourth and final chapter.

2

GENETIC EXPLANATION

In the previous chapter I quoted Richard Dawkins affirming his unqualified belief in 'the almost limitless power of the Darwinian principle'. This is an affirmation that has met with equally unqualified denial from other quarters. In 1994 David Stove, a respected philosopher writing in a professional journal, remarked:

> Most educated people nowadays, I believe, think of themselves as Darwinians. If they do, however, it can only be from ignorance: from not knowing enough about what Darwinism says. For Darwinism says many things, especially about our species, which are too obviously false to be believed by any educated person; or at least by an educated person who retains any capacity at all for critical thought on the subject of Darwinism.
>
> (Stove 1994: 267)

'[T]oo obviously false to be believed by any educated person'. This is the sort of thing that is frequently said by the 'cultured despisers of religion' (a memorable phrase coined by the theologian Friedrich Schleiermacher), of whom, as a matter of fact, Dawkins is one. Since Darwin is an authority frequently invoked by some of these modern despisers, it is intriguing to find it said of science rather than religion itself, and in one of its most prestigious and influential branches – Darwinian biology. Indeed, in another article Stove expressly describes Darwinism as a 'new religion'. What exactly might be meant by this, and whether it could be true, are topics to be returned to, but the charge itself draws attention to the fact that the most high-profile dispute about Darwinism has been its religious implications. I am referring, of course, to the fierce controversy between 'the theory of evolution' and 'creationism'.

Evolution and creationism

The most celebrated episode in the history of this dispute is the Scopes 'monkey' trial which took place in Mississippi in 1925. John Thomas Scopes, a high school teacher in Dayton, Tennessee, was prosecuted for having violated the state's recently passed law banning the teaching of human evolution in schools. The prosecution, led by the celebrated William Jennings Bryan, won its case, but the more lasting effect of the affair was to cast a poor light on creationism, partly because Scopes had let himself be prosecuted and partly because his defence lawyer, Clarence Darrow, was highly successful in exposing the intellectual poverty of the literalists as represented by Bryan. The anti-evolutionary law did not last long.

For the following few decades creationism appeared to be dead or at least dying. But somewhat surprisingly perhaps, it underwent a revival in the 1960s sufficient for the establishment of a Creation Science Research Center in California, and for more state laws to be passed in Arkansas and Louisiana. There had, however, been a significant change in emphasis. Now the creationists did not seek to have the teaching of evolution banned, but only to secure parity for the teaching of 'creation science'. Moreover they argued their case on the grounds of intellectual and scientific freedom rather than biblical revelation. After another celebrated court case, this time in a Federal Court, the Arkansas law was struck down.

In its second phase creationism played down the biblical interpretation on which it rested, though the Genesis Flood was extensively appealed to as a plausible hypothesis in its explanation of geological evidence. But the fact is that the motivating basis was still biblical literalism, an approach to the Bible that finds little support in the Western intellectual world of today, even (perhaps especially) among Christian biblical scholars. There are, certainly, people of considerable intelligence who are willing to defend creationism as a respectable alternative to evolution (Phillip Johnson, author of *Darwin on Trial*, is a notable example) and, like other intelligent proponents of indefensible positions, they often do so with impressive ingenuity. I shall not be concerned to examine their 'arguments', however, since these must sooner or later involve denying or explaining away an overwhelming body of empirical evidence to the contrary, and this simply cannot be done. Every serious investigation of the natural world serves to confirm rather than to refute Darwin's fundamental contention – that the indefinitely many species of plants and animals we find in nature have evolved over an enormously long period. Indeed, a significant body of scientific opinion had already come round to an evolutionary view of natural history before ever Darwin wrote *The Origin of Species* (see Ruse 1996: Ch. 3).

But a flaw in creationism even greater than its attachment to indefensible literalism and its disregard for empirical evidence, was the lack of any positive alternative. Despite its pretensions as a rival scientific theory, creationism is best characterized in purely negative terms – as anti-evolutionary – because there has long been a serious failure among creationists to agree on their theory of creation. This is why the term 'creation science' is misleading; the enduring division between strict creationists and progressive creationists is a radical one (see Numbers in Ruse 1996 Ch. 15; Miller 1999: 163–4). There simply is no single such science.

It is notable that, with the possible exception of Fred Hoyle, creationism has never had the support of any authoritative figures in the world of physics or biology. Its most celebrated proponents have been high school teachers, amateurs and (in Johnson) a lawyer. As a result among serious scientists it is widely, and rightly, regarded as wholly indefensible, a non-starter in fact. At the same time, despite a common assumption to the contrary, the rejection of creationism does not automatically imply the vindication of Darwinism. Indeed, it is extremely important to see that the closed mind of the convinced creationist is not infrequently replicated in the closed mind of the convinced Darwinian, and the technique of tarring those who raise important objections and difficulties for Darwinism with the same brush as the creationist is a favourite one among its protagonists. This is one of the reasons why Stove thinks of Darwinism as a new religion, as a dogma that drives the facts rather than a theory that is determined by them. This is not, in my view, an accurate characterization of religion *per se*, though it is certainly true that religious believers are often dogmatists. But dogmatism is an attitude into which scientists can also fall, and a danger to be especially avoided in considering the evolution/creationism debate. We need both to accept the indefensibility of creationism and to keep an open mind on the issues that divide Darwinism and its critics. In order to do so it is necessary to identify separate points of contention so that we can consider them in isolation from each other.

There are different versions of anything called 'Darwinism', but three distinguishable elements are specially important. The first of these is the relatively simple, factual claim that evolution has taken place, that plants and animals have evolved over many millions of years. The second is the more ambitious, more theoretical contention that this evolution is the result of two interacting factors – random mutation and natural selection. The third is the further proposition that *natura non facit saltum* – nature does not make leaps – which is to say that the process of evolution has been gradual, a long sequence of tiny changes. As Dawkins puts it, 'Wherever in the universe adaptive complexity shall be found, it will have come into being gradually through a series of small alterations, never through large and sudden increments in adaptive complexity' (Dawkins in Ruse 1996: 210). To these three elements, modern Darwinism adds a crucial further dimension of which Darwin himself was unaware – the genetic basis of life. Modern Darwinists such as Dawkins and and the philosopher Daniel Dennett hold that the ultimate explanation of the vast range of plant and animal life that we find on earth lies in the evolution (in accordance with Darwinian principles) of genes and genotypes. By their account the Darwinian explanation in terms of random mutation and natural selection is most powerfully at work with respect to genes rather than whole animals. To quote Dawkins again, 'we, and all other animals, are machines created by our genes. Like successful Chicago gangsters, our genes have survived, in some cases for millions of years, in a highly competitive world' (Dawkins 1976: 2).

The point of distinguishing these three separately identifiable elements, and the alternative levels at which they can be invoked (phenotype or genotype), is this. It is possible to accept one of these claims without accepting all of them, and/or to accept Darwinian explanations at one level but not at another. The three key propositions I have identified are not logically interconnected and there is no radical inconsistency in asserting the first and denying the third. It can even be held, consistently, that one of these propositions – the fact of evolution – is manifestly borne out by the evidence adduced to support it, while also holding that another – the principle of gradualism – equally manifestly is not. As we shall see, this position – that evolution is a fact but gradualism is false – is one that some serious contributors to contemporary scientific debate actually endorse.

It is also worth noting, for the purposes of analysis, that these three claims are not of the same kind. The first – the claim about evolution – though often held to be paradigmatic of a 'scientific' theory, is more accurately described as 'historical'. It is a claim, that is to say, about the past and how it has been. To invoke again some older terminology, it is a claim about natural history rather than natural philosophy, and simply describes an immensely long course of events which, so to speak, God might have watched unfold. By contrast, the second of these three elements is more properly thought of as 'scientific' in that it identifies an explanatory mechanism rather than an historical sequence – the interaction of random mutation and natural selection – a mechanism that is alleged to operate universally and without respect to particular times. The precise status of the third proposition – that the process of evolution has been gradual – is harder to determine. Is this also simply a claim about what has happened in the past that could, in principle though not in practice, have been observed? Or is it a necessary concomitant of the third, an implication that must hold if the explanatory mechanism of random mutation/natural selection is to be genuinely universal?

The significance of these remarks will emerge in due course. For the moment, however, I want only to observe that the fatal flaw in creationism is that it denies the first of the three elements in Darwinism, or more precisely it denies an important part of it – the time-scale. It is this that makes it indefensible. Before the idea of evolution took hold, understanding of the natural history of the world was to a great extent dominated by a literalist interpretation of the Book of Genesis, and people thought the earth (the universe indeed) to be a few thousand years old. The seventeenth-century Irish Bishop James Ussher, using the genealogies that are to be found in the Bible, calculated that the world had been created in 4004 BC and was thus roughly 6000 years old. This strikes us now as somewhat ludicrous, though in this as in all things, hindsight is wonderful. In fact, by the standards of any time Ussher was a man of immense learning, as is attested by his being accorded (most unusually) a state funeral in Westminster Abbey on the strength of his scholarly reputation. Radically wrong though he turned out to be, Ussher's assessment was the attempt of a serious intellect to calculate the age of the world on a reliable basis. By contrast, in sticking to the figure of 6000 years (or, on the part of some, the slightly vaguer one of 'under 10,000') modern creationists are wilfully ignoring an increased evidence base acknowledged by Christian scholars just like Ussher, several decades before Darwin.

Some creationists allow that the universe and the earth are much older than this, and hold to 6000 years as the length of time during which there has been life, or human life. Others opt for 10,000 years (see Ruse 1996, Ch. 16, a summary of Morris 1974). This uncertainty again reveals that creationists have no one positive theory, but in the light of the evidence, either figure is ridiculously wide of the mark, even for the biosphere, and when extended to include the earth and the universe it is even more absurd. Modern biology cannot endorse such a short period, and geology must operate within a vastly greater temporal expanse as well. In fact, it is worth remembering that Darwin himself was brought to think that the earth must be far older than had hitherto been thought by geological evidence – fish fossils high in the Andes. Similarly (though it is a point some creationists claim to be able to accommodate) the investigations of astronomers and cosmologists have extended the time-scale for the existence of the universe far beyond anything previously thought possible. Only those perversely determined to discount the immense strides that the natural sciences have made in the past 150 years could cling to the notion that an age of 6000/10,000 years might be a candidate for serious consideration. Kenneth R. Miller, a scientist deeply sympathetic to traditional theology, concludes a careful survey of the 'young creationist' contention by quoting the geologist G. Brent Dalrymple. 'The creationists' "scientific" arguments for a young earth are absurd', and remarks simply, 'Case closed' (Miller 1999: 77).

Creationism is thus rightly to be wholly discounted as a contender in the fields of biology, geology and cosmology. But its principal fault lies in trying to dispute the age of the earth that science everywhere confirms. To infer from this that the other contentions of Darwinism are equally uncontentious is a mistake. It is with the real difficulties attending natural selection and gradualism that the next few sections are concerned.

Natural selection and 'the selfish gene'

One of the most vexed topics relating to the theory of evolution in general and to Darwinism in particular is the idea of the 'survival of the fittest'. The expression was actually coined by Herbert Spencer in his *Principles of Biology* (1865), but in the fifth edition of *The Origin of Species* Darwin commends it over his own phrase 'struggle for existence', the title of Chapter 3. Controversy arose because from the first, 'survival of the fittest' so easily prompted a popular conception that the whole of the natural world including that of human beings was dominated by 'the law of the jungle' – an essentially normative idea that some regarded with fear and loathing while others took it up with enthusiasm. Now no doubt this was in large part an error, a misunderstanding of Darwin's contention, but if it was, it is one that has been re-invigorated by the application of Darwinism at genetic level. In particular, and most famously, it appeared to receive a considerable boost from Richard Dawkins's hugely successful book *The Selfish Gene* (1976) which has been translated into twenty languages and has sold over a million copies. On the surface, at any rate, Dawkins seems to embrace and endorse something very like the popular idea of 'survival of the fittest'. '[O]ur genes' he tells us, 'have survived, in some cases for millions of years, in a highly competitive world. This entitles us to expect certain qualities in our genes. I shall argue that a predominant quality to be expected in a successful gene is ruthless selfishness.' Moreover, 'this gene selfishness will usually give rise to selfishness in individual behaviour' (p. 2).

This contention led to a rather extraordinary exchange in the journal *Philosophy* where Dawkins was attacked vigorously, some would say intemperately, by the philosopher Mary Midgley, to whom Dawkins then replied in equally severe terms (see Midgley 1979; Dawkins 1981). Midgley, as it seems to me, took Dawkins too literally. Again and again he makes the point that when we take 'the licence of talking about genes as if they had conscious aims' we do so 'always reassuring ourselves that we could translate our sloppy language back into respectable terms if we wanted to' (Dawkins 1976: 88). Midgley's mistake, if that is what it was, is understandable, however, because Dawkins is himself inclined to draw inferences, especially about human behaviour, which suggest that the language of genetic selfishness and altruism is more than metaphor (e.g. 'Let us try to *teach* generosity and altruism, because we are born selfish.' Dawkins 1976: 3; emphasis in original). Furthermore, it is his use of these and other metaphors that won for his book the enormous attention it received. But we can, I think, restate the central biological ideas he expounds in a way that allows us to ask independently about their ethical and other implications. To do so it will be necessary to supply some elementary information about the genetic basis of plant and animal life for those who are not already familiar with it. (For a more detailed, and pellucid, account, see Russo and Cove 1998, to whom this short summary owes a great deal. Readers who already know the biological basics can skip the next few paragraphs.)

The bodies of all living things are composed of cells, many millions of them. Human beings are made up of an enormous number of cells – 10,000,000,000,000 or 10^{13}. Each cell is composed of thousands of kinds of molecules, which in their turn are made up of atoms. Atoms are commonly represented as miniature solar systems, in which the nucleus is the sun around which one or more electrons circle. The electrons on the edge of this system can interact with those on the edge of other systems, and it is by means of this interaction that atoms are built up into molecules. In the normal course of things molecules do not react with each other, but various factors such as the impact of heat may make them do so. In cells these reactions are induced by enzymes, a special kind of protein. Proteins are large molecules comprising strings of 20 types of amino acid, each with a distinct residue or combination of atoms.

All enzymes are proteins, but not all proteins are enzymes. Whereas the role of enzymes is to bring about interactions, the other proteins in cells have a variety of important functions, and it is these functions that determine the different types of cells of which the body is composed – blood cells, skin cells, nerve cells and so on. The proteins in a cell are synthesized by something called a ribosome. The ribosome is instructed as to what proteins to make by DNA (deoxyribonucleic acid), a long string of material contained within the nucleus of the cell. Because of its enormous length, which has to be packed within the tiny nucleus, DNA is structured into chromosomes. Human beings have 23 chromosomes, and each cell contains two copies.

This string of DNA is composed of a set of sequences of just four complex chemical compounds, the 'bases' of which all living things are composed, commonly referred to as A, C, G and T. The order of these sequences is crucial. It is in this order that we encounter genes, because a gene may be defined as a stretch of DNA that determines the synthesis (or manufacture) of a protein. Each of these stretches is marked by a 'start' and a 'stop' sign, both of which also take the form of combinations of the four bases. As Watson and Crick famously discovered in 1953, DNA has a double helix structure, two strands wound around each other. These two strands can separate and this is what makes replication of plants and animals possible. Since the four bases can only match up in accordance with certain 'pairing rules', when the strands are separated, each can determine the sequence of bases in a new and matching strand. If we know the sequence in one strand, we can predict the sequence of the other.

This enormously long sequence of bases with its start and stop signs is made even longer because of blocks of 'junk DNA' called introns, which often make up more of the sequence than the genetically significant DNA (in extreme cases as much as 99 per cent), and whose function (if it has one) is unknown. Taken together the genes into which DNA is divided constitute the genome, and it is the genome of each kind of entity that determines the kind of entity it is – the genotype. In turn the genotype determines the phenotype, which is to say, the physical entity with all its characteristics of size, shape, colour and so on. The human genome is made up of the sum of all the bases of our 23 chromosomes, and this is a huge number. Russo and Cove illustrate it in this way. 'Consider a book . . . with 3000 letters a page; we will need one million pages to write the whole list of the bases of our DNA. If we write the DNA sequence in books of 200 pages, that makes 5000 books, a very respectable encyclopaedia' (Russo and Cove 1998: 43). It is a measure of the speed with which genetics is advancing that though their book was published as recently as 1995, in answer to the question

'Can we write this encyclopaedia?' they could only say 'in principle, yes', and further predicted that the task would be accomplished in 2003. In fact, as we know, the task was completed in August 2000, three years ahead of schedule. (The significance of this accomplishment is a subject we will return to in a later chapter.)

One further point needs to be observed before the basic biological picture has been put in place. The replication of DNA, and hence of genes, is not infallible. In the copying of such vast sequences of letters, occasional mistakes occur. As a proportion, the number of these 'mistakes' is tiny, but the scale of the numbers is such that they are significant. Moreover, since DNA is 'read' in groups of only three letters – AGT, ACT, CTG, CCT, etc. – a mistake in just one letter can have considerable impact. If, for instance, the second letter of this sequence were omitted, it would read ATA, CTC, TGC, CT – a quite different sequence. This is what gives rise both to genetic variation and to defective genes. It is the process of 'random mutation' crucial to Darwinian biology, but at a genetic level. And this brings us back to Dawkins.

Darwin's original conception of natural selection was based on the idea of a perpetual competition among living things for survival. In this competition, random mutation can generate tiny advantages in the struggle to survive, and as a result any creature that is the beneficiary of such an advantage is likely to live longer. Being likely to live longer, it is likely to have more offspring, and these offspring will in turn inherit that advantage. Given a long enough period of time, such advantages add up to highly distinctive forms and features, better adapted to the conditions of survival, or more accurately to certain sets of conditions.

In the light of modern genetics it seems plausible to hold that the driving force behind such changes and consequent advantages is genetic modification, and indeed what we know of DNA supplies a convincing locus for them – tiny, occasional, mis-replications.

Dawkins often writes (misleadingly) as though genes were directly involved in the struggle for survival, but this is not what he means. Rather, genetic changes issue in phenotype changes. It is the phenotypes – whole animals, plants and so on – that are struggling to survive and to propagate. But the success with which they do this depends upon their genetic composition, and it is this genetic composition that their descendants inherit and which better equips those descendants for survival in their turn. What survives over the long term, though, is the genetic structure. The life of a phenotype is limited; it does not survive for long. What do survive, in some cases for millions of years, according to Dawkins, are the genes which, as he says (again somewhat misleadingly) 'jump' from phenotype to phenotype. 'Survival of the fittest', accordingly, must mean 'survival of the fittest genes'.

> The genes are the immortals, or rather, they are defined as genetic entities that come close to deserving the title. We, the individual survival machines in the world, can expect to live a few more decades. But the genes in the world have an expectation of life that must be measured not in decades but in thousands and millions of years.
>
> In sexually producing species, the individual is too large and too temporary a genetic unit to qualify as a significant unit of natural selection.
>
> (Dawkins 1976: 34)

Dawkins identifies three properties as being crucial to survival – longevity, fecundity and fidelity. The longer living a thing is, the more it replicates itself, and the more faithful to the original these replications are, the more it is likely to survive into succeeding generations. Fidelity is a property that we may attribute to the gene itself. Longevity and fecundity are properties of the body in which the gene has made its 'survival machine'. So far, so good. But why describe genes as 'selfish'? Once again the language is misleading. Since genes have no desires, they cannot have selfish desires. In fact, Dawkins's conception of selfishness is a technical one, as is its opposite, altruism.

> An entity, such as a baboon, is said to be altruistic if it behaves in such a way as to increase another such entity's welfare at the expense of its own. Selfish behaviour has exactly the opposite effect. 'Welfare' is defined as 'chances of survival', even if the effect on actual life and death prospects is so small as to *seem* negligible.
>
> (p. 4; emphasis in original)

Carefully considered, these definitions come close to being tautological. Since the genes that we detect today, whether in baboons or in babies, are those that have survived, it is evident that they must have behaved (or more accurately caused phenotype behaviour) in ways that resulted in their surviving, and hence in ways that increased their chances of survival. Consequently, they must have been 'selfish' as Dawkins defines it. The tautological character of these definitions is not necessarily objectionable and I do not mean to make much of it except to observe that 'selfishness' so defined has none of the *frisson* that many have found in the central thesis of *The Selfish Gene*, and that Dawkins's own talk of 'ruthlessness' is out of place. The gene does what it does (unless, as in the case of a lot of genetic material, it does nothing), on its own behalf and through the medium of the phenotype, its 'survival machine' and what it does may well be describable as 'selfish' in this technical sense. But it does not do what it does 'ruthlessly' as opposed to doing the same thing in some other way. To this extent, Midgley's complaints about Dawkins are not without foundation.

However, even when any 'sloppy' language has been purged, there is still an interesting and important thesis being advanced. And the central point to be considered about this thesis is just how far the genetic version of Darwinism takes us by way of explanation.

To address this issue properly, one more important element in the account of natural selection and the survival of the fittest must be recorded. This is an idea used by Dawkins but originating with another well-known and influential theorist, John Maynard Smith. It is to Maynard Smith that we owe the concept of an 'evolutionarily stable strategy' (ESS). This idea arises from an interesting and useful application of game theory to genetics. Game theory is a branch of rational action theory. It assumes that as far as possible rational agents will seek to promote their own self-interest, and it tries to devise action strategies and policies that will enable them to do so under conditions of uncertainty or ignorance, that is to say, in conditions where human agents either do not know or cannot be certain what other agents will do. Precisely because game theory assumes a self-interested motive and is concerned with action under conditions of ignorance, it is easily adapted to the struggle for survival in the natural world where it is also assumed that the relevant entities are driven by self-interest (the need to survive) and, being non-deliberative, are unaffected by conscious motives. This application of the game theoretic picture is one in which the fortunes of entities with certain sorts of *dispositions* are compared. These dispositions are not the outcome of rational reflection. They are to be thought of as genetically in-built tendencies to act in accordance with a certain strategy – for example to attack rivals and fight to the death, say, as opposed to running away from rivals, or rapidly conceding defeat.

An ESS is one which, if pursued by most members of a population, cannot be bettered by an alternative. The qualification 'most members' is very important. It might seem obvious for instance that in any clash between a hawk-like (i.e. aggressive) strategy and a dove-like (i.e. non-aggressive) strategy the hawk-like strategy is bound to win out. But in fact, when we think these strategies through as natural dispositions to behaviour working themselves out over time, it often happens that the obvious strategy is not one that could be pursued by the majority of a population without wholesale destruction (i.e. zero survival chances). The natural strategies that really do increase the chances of survival are not obvious, and accordingly there are interesting insights into ecological balances to be gained from thinking about these things game theoretically.

The combination of genetics, game theory and evolution – modern (or neo-) Darwinism in short – is a powerful one. But just how powerful is it? Dawkins, as we have seen, believes that its explanatory power is unlimited and universal; there is nothing that it cannot explain. This is an ambitious claim. Is it true? I shall consider the explanatory power of Darwinism in four connections – the zoological, the biochemical, the psychological and the cultural. But to begin with, I want to note a more general difficulty that Darwinism encounters in all its versions and applications. This is the definition of a key concept – the survival of the fittest.

Survival of the fittest

It is a long-standing criticism of the concept of 'survival of the fittest' that it is circular. That is to say, in using it to explain the existence of the things around us, there is a real danger of supplanting the genuinely empirical with the *a priori*. How fit does the fittest have to be? Does it just mean *fit enough*? Since everything we see around us by way of life forms has, obviously, survived, then it must by definition have been *sufficiently* fitted to survive. Interpreted in this way, the explanatory principle of 'the survival of the fittest' seems to collapse into 'the survival of the surviving', and while the original version might be thought to be a principle of substance, its interpretation along these lines is evidently tautological. How is this collapse into circularity to be avoided? Someone coming new to the subject might reasonably imagine that this is a question to which ready answers have long ago been supplied. Surprisingly, perhaps, this does not appear to be true. There is no widely acknowledged conception of fitness that clearly and unmistakably avoids this difficulty, and the fact that this is so is testimony to how difficult it is to define 'fitness' in a non-circular way.

By way of illustration, consider the application of the concept at the level of the genotype. It seems plausible to suppose that a useful conception of evolutionary fitness would lie in the efficiency with which a given type of plant or animal uses the energy (or food) available to it in the ecological niche it occupies. If it uses the food source available as efficiently as possible, this would explain why it has survived in this particular niche where other, less energy-efficient genotypes have not. This interpretation of 'survival of the fittest' does avoid circularity, and renders it empirically substantial. Unfortunately, this sort of calculation (in so far as we are able to make it) does not seem to square with the facts. There is no special reason to think that energy-use efficiency is the crucial factor in determining the survival of a particular phenotype or genotype. This is because of the existence of an alternative though equally influential factor – relative fecundity. To put the point simply: an animal that makes relatively poor use of the energy in its environment and is for that reason short lived, may nevertheless reproduce in large numbers during the short life it enjoys. By contrast, an animal that is long lived because it does make first-rate use of the food supply may breed rarely and slowly. Where this is the case, there is good reason to think that the genes of the less 'fit' but more fecund animal are more likely to survive. And the same point can be made about a third important factor – reproductive fidelity. A creature that is both efficient and fecund will propagate its genes less successfully should it have a tendency to less faithful reproduction.

In any case, even if we were to settle on a useful and widely accepted concept of fitness (some combination of all three factors alluded to, perhaps) we would still face further considerable difficulty. 'Fitness', however we characterize it, has to be fitness in some environment or other. It follows that radical changes in the environment can undermine a genetic inheritance that has, hitherto, been supremely well suited to prevailing conditions. For example, if it is true (as often alleged) that the dinosaurs were wiped out by radical climatic change brought about by a long series of volcanic explosions, then it is also true that the explanation does not lie exclusively with genetic fitness. The dinosaurs were perfectly fitted to one environment, and wholly unfitted to another. The reason why they do not exist, therefore, cannot be accounted for exclusively in terms of 'fitness'. Crucial to the explanation is a factor that has nothing whatever to do with genetics, namely geological and climatic history.

Now this second point, conceptually speaking, is of the greatest importance. What it demonstrates is that when it comes to explaining the living world, genetic mutation and biological selection can only take us so far. The contingencies of the non-genetic, non-biological environment are crucially important also, and to this extent, if to no other (*pace* Dawkins), the explanatory power of the Darwinian principle is *not* limitless. On the contrary, if the case of the dinosaurs is anything to go by, Darwinian explanation fairly speedily reaches its limit. But, though conceptually important, this point is unlikely to persuade many readers of the deficiencies of Darwinism. Certainly, they will say, the brute contingencies of geological history and climatic change enter the explanation. But what

Darwinism reveals is an underlying mechanism that is *not* brutely contingent, an explanatory principle deeply built into the nature of things, including the nature of human beings. The admitted impact of volcanoes, ice ages and the like does nothing to alter this, and hence does nothing to alter its power or importance.

I am myself persuaded by this rejoinder, and hence persuaded that Darwinism is not much the poorer for having to admit the contributions of geology and climate to the world we find around us. Nor, indeed, am I inclined to make much of the first objection I raised – the absence of any clear, substantial and agreed conception of fitness. This is because, faced with uncertainty about the competing claims of fitness, fecundity and fidelity, a convinced Darwinian can take refuge in further abstraction. The key conception which Darwinism employs (in addition to that of random mutation) is not so much the idea of *fitness for* as that of *advantage in* the struggle for survival. Darwinism can still be said to supply us with a plausible and powerful explanatory framework, provided only that it can make good the claim that (barring accidents of geology and climate) survival is the outcome of accumulated advantages vis-à-vis other species arising from genetic mutation.

Altruism, homosexuality and sterility

Having disarmed rather than disposed of these general difficulties in the concept of 'survival of the fittest', we can turn to another very familiar line of objection to evolutionary biology – that there are important aspects of observed and recurrent animal behaviour which neither Darwin's natural selection nor its descendant the selfish gene can explain. Three of these are regularly repeated as constituting important obstacles, namely altruism, homosexuality and sterility. According to E. O. Wilson, 'The central theoretical problem of sociobiology [is]: how can altruism, which by definition reduces person fitness, possibly evolve by natural selection?' (Wilson 1975: 3). Altruism in this context means any behaviour on the part of an individual creature that is such as to increase another individual's welfare at the expense of its own. Apparent instances of this type of behaviour are not far to seek, it seems. We do not need to reach for the high-level behaviour of heroic self-sacrifice on the part of human beings in times of war. More modestly, there are animals and birds that will give out warning signals to the other members of their herd or flock at the approach of a predator. While this may enable their confrères to get away, it draws greater attention to themselves and their position, and as a result makes it more likely that they will be caught and devoured. How is it, if the evolutionary story is correct, that the genes which prompt such warning behaviour have not been wiped out long ago? At first glance it seems obvious that the survival rate of those individuals with this tendency is significantly *lower*, and hence that the initial mutation which brought it about would not have been favoured in the struggle for survival.

More damaging yet, it would appear, is behaviour that actually mitigates against reproduction. The plainest instance, perhaps, is homosexuality. The tendency to engage in homosexual, i.e. non-reproductive, sexual acts is to be found in a wide range of animals, and not just human beings. But, once again, on the surface this is puzzling. If, as genetic biology holds, the key notion in explaining the origin and variety of species is random genetic modification that results in greater genetic fitness to survive, this can hardly explain how a genetic disposition *not* to reproduce persists, as it obviously does in a great many species. The genes, whichever they may be, that lie at the base of homosexuality can hardly replicate themselves successfully, since they don't replicate themselves at all!

The most dramatic counter-instance, of course, is the persistence of sterile creatures, notably ants and other insects. Darwin himself regarded this as a special difficulty, 'which at first appeared to me insuperable, and actually fatal to my whole theory. I allude to the neuters or sterile females in insect communities: for these neuters often differ widely in instinct and structure from both the males and fertile females, and yet, from being fertile, they cannot propagate their kind' (Darwin 1996: 192). His (initial) perplexity is understandable. How can the theory of the selfish gene explain the continued existence of creatures that never reproduce because they cannot? Surely the moment genetic mutation takes this unhappy turn, the type of being in question is 100 per cent genetically *unfit* to survive. But there are such creatures.

In response to these sorts of case, some biologists have argued that what is at issue is not the survival of the individual, but the survival of the group or species – an idea that is sometimes known as 'group-selection theory'. Konrad Lorenz, for instance, speaks of the species-preserving functions of aggressive behaviour. There is an interpretation of this claim that is not wholly incorrect, but it is seriously misleading, and such a response marks an important division between evolutionary biology in some of its versions and the genetic version which now holds sway. Dawkins, for one, is quite clear that group selection as an explanation of the altruistic behaviour of individual animals is mistaken on a number of counts. And it is easy to see why it does not accord with the genetic conception. Genes are embodied in individuals, not in groups. Accordingly, as he says, if 'the best way to look at evolution is in terms of selection occurring at the lowest level of all' (Dawkins 1976: 11), i.e. at the genetic level, then the relevant 'survival machine' is and can only be the individual phenotype, not the group.

In any case, evolutionary biology has no need of group-selection theories because on closer inspection the phenomena of altruism, homosexuality and sterility are not so very troublesome after all.

This is why I have carefully qualified my explication of these supposed counter-instances with phrases such as 'it seems' and 'it would appear'. To understand how a theory such as Dawkins's accommodates them it is worth noting at the outset that nothing in it implies that there is 'a' gene for altruism or homosexuality, or even for sterility. We tend to be misled on this point by generalizing illicitly from the much cited instance of brown and blue eyes, for which there is an identifiable gene. But both behaviour and physiology are often determined by a combination of genes, rather than just by one. It is a point of very considerable importance to which we will return at greater length in another chapter. For the moment, however, it is enough just to note it.

Actually, despite its tendency to mislead, the example of blue and brown eyes is quite instructive in this context. Since an individual cannot have both brown and blue eyes, where alternative genes are embodied in a single phenotype (one from the father, the other from the mother), then one gene will be dominant and the other recessive. But it may still be passed on to progeny and in the right combination become dominant in later generations. So this is a possible scenario. The genetic basis of altruistic or homosexual behaviour is in the main a recessive gene, which emerges from time to time within a general genetic inheritance that is 'fit'. As a consequence, there will be some individuals with seemingly self-destructive genes, but these will be genes that are for the most part recessive.

It might be thought that this hardly answers the problem, and in itself, probably it does not. A further point will close the gap, however. It is possible, and in the cases we are considering highly likely, that the presence of this trait increases the survival chances of the total genetic package, so to speak. This is the limited truth that is to be found in the talk of 'group-selection'. E. O. Wilson makes this point with respect to homosexuality. There is, he says,

> a strong possibility that homosexuality is normal in a biological sense, that is a distinctive beneficient. . . . Homosexual behaviour is common in other animals, from insects to mammals, but it finds its fullest expression as an alternative to heterosexuality in the most intelligent primates, including rhesus macaques, baboons and chimpanzees. . . . This special homophile property may hold the key to the biological significance of human homosexuality. Homosexuality is above all a form of bonding. It is consistent with the greater part of heterosexual behaviour as a device that cements relationships. The predisposition to be homophile could have a genetic basis, and the gene might have spread in the early hunter-gatherer societies because of the advantage they conveyed to those who carried them.
>
> (Wilson 1995: 138)

The principal point is that there is more to the survival of higher animals than mere replication after the fashion of bacteria, and the genes of creatures that have strong social inclinations will have a greater chance of persisting. Homophilia, of which homosexuality is just one – if the most obvious – expression, may both reflect and serve these inclinations.

In a different way, the same point is to be made about sterility. The biology is more complicated here, but there is reason to think that the role of sterile worker bees within the organization of the hive is such that the survival probability of the genetic structures that give rise to them is increased. It is worth quoting Darwin's own solution at this point.

> How the workers have been rendered sterile is a difficulty; but not much greater than that of any other striking

> modification of structure; for if it can be shown that some insects and other articulate animals in a state of nature occasionally become sterile; and if such insects had been social, and it had been profitable to the community that a number should have been born capable of work, but incapable of procreation, I can see no very great difficulty in this being effected by natural selection. But I must pass over this preliminary difficulty. The great difficulty lies in the working ants differing widely from both the males and the fertile females in structure as in the shape of the thorax and in being destitute of wings and sometimes eyes, and in instinct.
>
> (Darwin 1996: 192)

> This difficulty, though appearing insuperable, is lessened, or, as I believe, disappears, when it is remembered that selection may be applied to the family, as well as to the individual, and may thus gain the desired end. . . . Thus I believe it has been with social insects: a slight modification of structure, or instinct, correlated with the sterile condition of certain members of the community, has been advantageous to the community: consequently the fertile males and females of the same community flourished and transmitted to their fertile offspring a tendency to produce sterile members having the same modification. And I believe that this process has been repeated, until that prodigious amount of difference between the fertile and the sterile females of the same species has been produced.
>
> (Darwin 1996: 193–4)

All three phenomena are thus consistent with the underlying theory of 'genetic fitness', and in order to appreciate this, we need only see that the relationships between different genes and the various behaviours they prompt are rather more complex than the initial interpretation of the supposed counter-instances allows. This sort of complexity is illustrated by another widely discussed example – sickle cell anaemia. Sickle cell anaemia (so called because it makes the red blood cells sickle shaped) is caused by a mutant recessive gene. Those who have a double copy of the gene will sicken and die at a fairly early age. Why does it still persist? Why has its dramatic impact on reducing the survival rates of the individuals who have it not caused it to disappear over time? The answer is that individuals who have a single copy of the same gene are more resistant to malaria, which explains why sickle cell anaemia is more common in Central and North Africa than in any other part of the world. The gene is thus more beneficial than detrimental to survival, despite a natural supposition to the contrary.

All these cases are interesting and important as potential empirical tests for the theory of evolution. But as I have been suggesting, and as Dawkins, Wilson and Darwin himself contend, closer inspection shows that they can be satisfactorily accounted for within the general framework of genetic explanation that modern evolutionary biology employs. The key conception is not that of group selection but of population genetics. Once we understand the way in which the gene pool operates not just with respect to individuals but within populations (not groups) of individuals, the difficulties they seem to present fade away. In short, provided we think about them in the right way, these intriguing phenomena do not constitute any real challenge to the Darwinian framework.

However, if the longstanding objections based on altruism, homosexuality and sterility can be dismissed, there is reason to think that a more serious scientific challenge has recently emerged elsewhere. Genetics is one of the most important new branches to develop in the comparatively recent history of science, and with dramatic consequences. But it is not the only one. No less important is biochemistry, if anything an even newer branch of science, and biochemistry has revealed to us many startling facts and phenomena at the microscopic interface between the chemical and the biological. It is to these biochemical challenges to genetic explanation that we now turn.

Irreducible complexity and the biochemical

A recent book that has drawn almost as much popular attention (especially in the United States) as *The Selfish Gene* is Michael J. Behe's *Darwin's Black Box*. The reception of this book in some quarters is a good illustration of the tendency among Darwinians to brand all criticism as 'creationist', matched, it must be said, by the tendency of some creationists to herald any criticism of Darwinism as support for their cause. But the subject with which Behe's book is concerned transcends the dispute between Darwinians and creationists (with whom he differs on several crucial points; see Miller 1999: 164). Commenting on a reply by Dawkins to the creationist Francis Hitching's appeal to the phenomenon of the extraordinary bombardier beetle, Behe writes:

> The problem with the above 'debate' is that both sides are talking past each other. One side [the creationists] gets its facts wrong; the other side merely corrects the facts. But the burden of the Darwinians is to answer two questions. First, what exactly *are* the stages of beetle evolution, in all their complex glory? Second, given these stages, how does Darwinism get us from one to the next?
>
> (Behe 1998: 34)

It is the second of these questions that is of special importance. What Behe raises doubts about is gradualism, and he rightly sees that doubts about its gradualist component are of very great importance in assessing the explanatory power of Darwinism as a whole. In this connection he appositely quotes Darwin himself:

> If it could be demonstrated that any complex organ existed which could not possibly have been formed by numerous, successive, slight modifications, my theory would absolutely break down.
>
> (*The Origin of Species*, sixth edition, p. 154; quoted in Behe 1998: 39)

Behe thinks that on a small scale, Darwin's theory has triumphed and is quite uncontroversial (another point of difference with the creationists). But he contends that at the level of biochemistry rather than biology (about which we know a great deal while Darwin knew virtually nothing), Darwinian gradualism faces insurmountable obstacles. The key concept on which his argument is based is that of 'irreducible complexity'.

> By *irreducibly complex* I mean a single system composed of several well-matched, interacting parts that contribute to the basic function, wherein the removal of any one of the parts causes the system to effectively cease functioning. An irreducibly complex system cannot be produced directly (that is, by continuously improving the initial function, which continues to work by the same mechanism) by slight, successive modifications of a precursor system, because any precursor to an irreducibly complex system that is missing a part is by definition nonfunctional. An irreducibly complex biological system, if there is such a

> thing, would be a powerful challenge to Darwinian evolution.
>
> (Behe 1998: 39; emphasis in original)

The larger part of the chapters that follow is devoted to showing that not only is there such a thing as an irreducibly complex biological system, but that irreducible complexity is everywhere in the biochemistry of life. Among the instances Behe explores in detail the most striking are those of the immune system and blood clotting. It is evident, I think, that a creature possessed of even a primitive immune system, and/or whose blood had the ability to clot, would enjoy enormous advantages in the struggle to survive over other creatures that did not. Moreover, assuming that such systems have a genetic base, then the genes of such creatures would be propagated, and hence perpetuated, far more readily and widely than those of creatures who had no defence against disease and/or who were haemophiliac. So at one level, the widespread distribution of such important biological mechanisms is confirmation of the Darwinian principle of survival of the fittest and in so far as we follow Dawkins's picture of phenotypes as essentially survival machines for genes we can think of these highly valuable phenotypical properties as means by which the genes which underlie them continue their existence from generation to generation. Even so there is the further question of how such systems could arise. It is on this issue, if Behe is correct, that Darwinism fails, and it fails because its commitment to gradualism – *natura non facit saltum* – prevents it from explaining the fact of such mechanisms. 'Biochemistry has . . . revealed a molecular world that stoutly resists explanation by the same theory so long applied at the level of the whole organism' (p. 173).

The central point in his argument can be made schematically. Suppose that a mechanism D has three component parts A, B and

C. It requires all three if it is to function; one or two will not do. But this means that the emergence of A and B, let us say, does not give their possessor part of the advantage that it will have when it possesses all three. The mechanism is an all or nothing affair. Consequently we cannot construe the emergence of D as the cumulative process of acquiring A, B and C over time and in accordance with natural selection, because though natural selection may massively favour the possessor of D it does not thereby favour the possessor of A and B. 'If A, B and C have no use other than as precursors to D, what advantage is it to an organism to make just A? Or, if it makes A, to make B?' (p. 151). 'I emphasize that natural selection, the engine of Darwinian evolution, only works if there is something to select – something that is useful *right now*, not in the future' (p. 95; emphasis in original).

In fact the difficulty for Darwinian explanation in some of the phenomena Behe discusses seems to be even worse than this.

> The regulation of AMP [a form of the nucleotide A] biosynthesis is a good example of the intricate mechanisms needed to keep the supply of biomolecules at the right level: not too much, not too little, and in the right ratio with related molecules. The problem for Darwinian gradualism is that cells would have no reason to develop regulatory mechanisms before the appearance of a new catalyst. But the appearance of a new, unregulated pathway, far from being a boon, would look like a genetic disease to the organism. This goes in spades for fragile ancient cells, putatively developing step by step, that would have little room for error. Cells would be crushed between the Scylla of unavailability and the Charybdis of regulation.
>
> (Behe 1998: 159)

In other words, some of the As, Bs and Cs that go to make up a mechanism, far from bringing survival advantages eventually leading to D, would, on their own, prove lethal.

In the face of such biochemical mechanisms Behe considers some of the outline gradualist explanations that have been hypothesized. He finds even the most plausible and promising of them deficient, and while I am not in a position to adjudicate on the scientific findings he invokes, to the 'intelligent layman' he makes out a very good case. However, as one of his most sympathetic and careful reviewers points out, Behe illicitly supposes that because some gradualist solutions do not work, *none* can, and this seems to be false.

> Behe's colossal mistake is that, in rejecting these possibilities, he concludes that no Darwinian solution remains. But one does. It is this: an irreducibly complex system can be built gradually by adding parts that, while initially just advantageous, become – because of later changes – essential. The logic is very simple. Some part (A) initially does some job (and not very well perhaps). Another part (B) later gets added because it helps A. This new part isn't essential, it merely improves things. But later on A (or something else) may change in such a way that B now becomes indispensible. This process continues as further parts get folded into the system. And at the end of the day, many parts may all be required.
>
> The point is there's no guarantee that improvements will remain mere improvements. Indeed because later changes build on previous ones, there's every reason to think that earlier refinements might become necessary. The transformation of air bladders into lungs that allowed

> animals to breathe atmospheric oxygen was initially just advantageous: such beasts could explore open niches – like dry land – that were unavailable to their lungless peers. But as evolution built on this adaptation (modifying limbs for walking, for instance), we grew thoroughly terrestrial and lungs, consequently, are no longer luxuries – they are essential. . . . [A]lthough this process is thoroughly Darwinian, we are often left with a system that is irreducibly complex.
>
> (Orr 1996/7: 2)

This Darwinian rejoinder, it should be added, is perhaps not as destructive as its author supposes. According to Behe, there are biochemical mechanisms – blood clotting being a specially striking one – for which gradualist explanations are not available. What Orr has shown, in my view, is that there is a Darwinian scenario of a kind Behe has not considered. It is, however, a 'how possibly' story, and what is needed is a 'how actually' account of the phenomena Behe describes at such length. Kenneth R. Miller, who cannot be accused of any anti-religious bias, claims that this has in fact been done convincingly, and has done so for blood clotting. After an extended review of some important papers, several (but not all) of which appeared after Behe's book was published in 1996, Miller concludes:

> Michael Behe's purported biochemical challenge rests on the assertion that Darwinian mechanisms are simply not adequate to explain the existence of complex biochemical machines. Not only is he wrong, he's wrong in the most spectacular way. The biochemical machines whose origins he finds so mysterious actually provide us with powerful and compelling examples of evolution in action. When we

go to the trouble to open that black box, we find out once again that Darwin got it right.

Remember the Darwin quotation that Behe used to justify his attack on evolution?

'If it could be demonstrated that any complex organ existed, which could not possibly have been formed by numerous, successive, slight modifications, my theory would absolutely break down.'

In the light of the biochemical systems we have just explored, it's too bad that Behe didn't go to the trouble to print the sentence that followed:

'But I can find no such case.'

Neither I would add, has Michael Behe.

(Miller 1999: 160–1)

However, Behe has another arrow in his quiver. This lies in the contention that Darwinism is silent with respect to the topic that many think to be the most fundamental of all – the origins of life. This issue would appear to be crucial partly because it is only life forms that have the power of replication that lies at the heart of Darwinian explanation, and partly because the emergence of life from non-life requires us to construct, if only in outline, the means by which it is possible. Of course, it is quite widely believed that this has been done, that the seemingly yawning gap between the organic and the inorganic was bridged by the experiments of Stanley Miller conducted in Chicago shortly after the Second World War. But according to Behe, Miller's original findings have not in fact been developed in the directions in which they were widely expected to be. By passing electricity through a mixture of gases and a pool of water Miller produced amino acids. At the time this was heralded as a crucial breach in the fundamental difference that had previously been thought to divide life from non-life. It also ushered in all the subsequent talk of a 'primeval soup' from which, thanks to lightning storms, life first emerged upon earth, an idea that Dawkins frequently invokes in *The Selfish Gene*. But whereas this was widely interpreted as the first step in a series of experiments that would finally reveal the evolutionary process by which life itself arose from the 'dead' earth of billions of years ago, it appears that none of the crucial subsequent steps have ever been forthcoming. Behe quotes Klaus Dose whom he describes as a prominent worker in the field:

> More than 30 years of experimentation on the origin of life in the fields of chemical and molecular evolution have led to a better perception of the immensity of the problem of the origin of life on earth rather than to its solution. At present all discussions on principal theories and experiments in the field either end in stalemate or in a confession of ignorance.
>
> (Dose 1988 quoted in Behe 1998: 168)

Certainly this much is true. Miller's experiments, whatever success we attribute to them, are relevant to the question of the actual origins of life only if we have reason to think that they replicate something like the conditions that prevailed 2.4 billion years ago when (some estimate) life began, and we have no reason to do so. Talk of primeval soups and electric storms is no more than conjecture, and moreover, conjecture driven largely by the supposed success of Miller's experiment. We really have no independent grounds for such a supposition. Even without such grounds, of course, the theoretical position would be different if life had in fact been created under laboratory conditions. While it might remain true that we continued in ignorance of the conditions actually prevailing at the origin of life, we would at least have some idea about how the emergence of the organic from the inorganic was possible. But importantly, nothing of the sort has happened. Kenneth R. Miller addresses this point.

> We could, if we wished, hold up the origin of life itself as an unexplained mystery. . . . Since neither I nor anyone else can yet present a detailed, step-by-step account of the origin of life from non-living matter, such an assertion would be safe from challenge – but only for the moment. . . . None of [the facts we currently know] proves that life originated purely of naturalistic causes, and therefore none of them proves that the first cell was not the direct, miraculous, intentional work of a Creator. At least not *yet*, and there's the danger.
>
> (Miller 1999: 276; emphasis in original)

Now it is important for present purposes to see that Miller's point is made in the context of the debate between science and religion. Rightly, in my view, he does not want to relegate religious belief to the gaps in scientific knowledge because he sees that this is a position of perpetual retreat. But if this is not our concern, if our interest is solely in the scope of evolutionary explanation, we have to concur with Behe that the origin of life has not yielded to evolutionary explanation. We must of course add 'Not yet', but this acknowledgement ought not to carry any implication that in time it is bound to. This is the supposition that the most convinced Darwinians seem to make. '[I]f there is no other generalization that can be made about Life all round the Universe. I am betting that it will always be recognizable as Darwinian life. The Darwinian Law . . . may be as universal as the great laws of physics' (Dawkins in Ruse 1996: 220). If this contention is not sheer dogma, it must rest on faith, which is one of the reasons why Stove and others have regarded Darwinism as 'A New Religion'.

In summary: the boldest version of Behe's case rests not just on the contention that Darwinian explanations of these intriguing biochemical mechanisms have not been forthcoming, but that they cannot even be formulated since the irreducible complexity of (many) biochemical mechanisms is not susceptible to gradualist explanation. This is a conceptual rather than a scientific argument and it may be (as Orr in effect argues) that complexity of the kind that Behe identifies is not in principle irreducible to evolutionary steps. However that may be, there remains the less ambitious contention that as a matter of fact, no such explanations have been forthcoming. Indeed a significant portion of Behe's book is devoted to an exhaustive survey of the relevant scientific literature, from which he concludes not only that there are no such explanations currently available, but that the very few outline sketches that some biologists have produced are deeply implausible, or question begging. Reputable scientists have disputed this, and evidence on this point may have changed since Behe's book was published. The issue under discussion here is, of course, a scientific one. At some point, those like me (and most readers of this book I imagine), who are not themselves expert in the biochemistry must trust to those who are. We should conclude, I think, that Behe overstates his case and that, though there may be biochemical phenomena which we cannot at present explain in evolutionary terms, this is a result of our current ignorance and not of their 'irreducible complexity'. The best case he makes is for thinking that in explaining the origins of life itself, the genetic version of evolutionary biology faces a challenge it may not be able to overcome. The biochemical, then, is one sphere that may set limits to the supposedly 'limitless power of the Darwinian principle'. Are there any more? Arguably, psychology is another.

Sociobiology and evolutionary psychology

Human beings have a psychology as well as a physiology. In addition to the ways in which they procreate, gestate, take nourishment, grow, move and age, they exhibit a mentality. This is not just a matter of patterns of behaviour, such as ants and termites also exhibit. People are creatures of thoughts, feelings and desires, and the masters of languages in which these thoughts, feelings and desires are expressed and communicated. Can genetic explanation stretch to explaining these important features of their existence also?

Among the best known of those who believe it can is the Harvard scientist E. O. Wilson. Wilson it was who invented the term 'sociobiology' in a long book with that title, first published in 1975. As the name suggests, sociobiology is the attempt to use the methods and findings of biology in the understanding of social phenomena. Wilson is an entomologist by profession, someone who studies insects, and of course many insects live in highly complex social formations. Given that, like insects, human beings, whatever else they may be, are members of an evolved animal species, in the abstract at any rate, the project seems a not implausible one. Following the publication of *Sociobiology*, Wilson was roundly criticized for ignoring all those studies of human behaviour and social formation that have been undertaken over a long period by economists, sociologists and especially anthropologists. In a subsequent, shorter volume, entitled *On Human Nature*, Wilson attempted to remedy this deficiency, and it is with some of the views he expounds in this later book that we are chiefly concerned here.

Wilson is not offering us an alternative to explanation in terms of genetic structures. On the contrary, he accepts the genetic basis of all life and its importance in explaining the huge variety of forms and behaviours with which the sciences are confronted. In *On*

Human Nature he sets out the same basic framework elaborated by Dawkins, Maynard Smith and so on.

> The heart of the genetic hypothesis is the proposition, derived in a straight line from neo-Darwinian evolutionary theory, that the traits of human nature were adaptive during the time that the human species evolved and that genes consequently spread through the population that predisposed their carriers to develop those traits.
>
> (Wilson 1995: 31)

But Wilson seeks to extend the reach of this hypothesis into the explanation of phenomena more usually associated with the social sciences. So when he turns his attention to sex, for example, his principal concern is with patterns of sexual (and associated) behaviour rather than with the mechanics of reproduction. And he devotes an entire chapter to a topic Dawkins only touches on – religion. I shall not be concerned directly with what he has to say about them here, however (though the question of religion will figure prominently in Chapter 4). For present purposes it is enough to observe that Wilson's aim is to give a Darwinian explanation of the social as well as physiological nature of human beings, and to do so in a way that is both in accordance with the fundamental principles of genetic biology, and at the same time draws upon the investigations of anthropology and related human studies.

This sociobiological project has been given greater refinement (and in some ways an even larger ambition) by extension to psychology. Evolutionary psychology, as its name suggests, is an even more recent attempt to apply the insights and methods of evolutionary theory to the study of the human mind. Whereas sociobiology brings social interaction within the scope of the Darwinian programme, evolutionary psychology aims to extend it to the typical subject matter of psychology – sensation, emotion, deliberation, belief and memory – and it does so by combining experimental psychology and anthropology. Indeed it would not be inaccurate to say that in some quarters this evolutionary turn has been heralded as the great white hope of psychology writ large, a development that will rescue it from the aridity of the post-Freudian neuro-physiology to which (some allege) psychology has fallen victim. Whether or not such hopes are well placed remains to be seen, but there is no denying that the ambition of evolutionary psychology in this regard has attracted a great deal of attention. One of the most prominent of its exponents is the experimental psychologist Steven Pinker, the title of whose book – *How the Mind Works* – reveals something of this ambition (an ambitiousness he acknowledges in the preface).

This extension is not entirely unexpected. Darwin himself thought the advent of an evolutionary understanding of nature would not be confined to the biological. 'Psychology will be based upon a new foundation', he writes at the end of *The Origin of Species*. Yet as Pinker notes, 'Darwin's prophecy has not been fulfilled. More than a century after he wrote these words, the study of mind is still mostly Darwin free, often defiantly so' (Pinker 1997: 22). The self-appointed task of evolutionary psychology, in short, is to remedy this state of affairs by advancing the psychological aspect of the explanatory programme Darwin initiated.

One key idea in so doing is that of 'reverse engineering'. If it is true that the complexity of mind and body we find in human beings today is the outcome of an enormously long process of evolution, then understanding that complexity will be a matter of reconstructing that process, reasoning backwards from what we find in human (and other) behaviour, to the evolutionary path by which it must have come about. The path by which it *must* have come about? The reference to necessity here raises a point of considerable consequence. It implies, if carelessly interpreted, that 'reverse engineering theory' explanations are largely a matter of deduction. But as I stressed at an earlier stage of the argument, the evolutionary path that human beings have trod is a matter of natural *history* – how things have been in times past as a matter of recordable fact – and this is not something that can be deduced. What may appear to be highly convincing deductions drawn from our knowledge of the brain (say), or for that matter from close entomological study of the social insects, are in reality only *hypotheses* about how things have been, and as hypotheses they are as good as and no better than the evidence against which they can be checked.

Of course, one thing that differentiates natural history from history as we usually think of it is its slowness; the natural as opposed to the political history of human beings is a story of change so gradual that it needs to be conceptually 'speeded up' (so to speak) if it is to be grasped at all. Even so, it is evident that if the idea of 'reverse engineering' is not to fall victim to a vicious apriorism of the kind to which (as we saw) the central Darwinian idea of the survival of the fittest has been prone, then at every crucial point empirical evidence must be supplied. And the question arises – where is this evidence to come from?

At this point we encounter an important difference between evolutionary biology and evolutionary psychology. Whereas, when it comes to the physiology of human beings and other animals, we can hope to trace the evolutionary path with the help of the fossil record – and have been able to do so with some success – the same body of evidence is not really available to us with respect to psychology; there are no mental fossils. Why does this matter? Well, suppose that we are engaged in the reverse engineering of some physiological feature – the emergence of precision hands, for instance (a development of considerable survival value, of course). Our 'reverse engineering' theory will locate the precision hand in an evolutionary sequence, and then we can use the fossil record to check whether that (roughly) is indeed the sequence in which it is to be found. But suppose that what interests us is the emergence of some intellectual trait or emotional state – the ability to count, or strong maternal feeling, say. In these cases, no less than in the physiological one, our reverse engineering theory, however plausible it may seem to be, needs to be checked in some way against 'what actually happened'. For instance, Pinker says 'Precision hands and precision intelligence co-evolved in the human lineage, and the fossil record shows that hands led the way' (p. 194). But the reference to the fossil record here is one-sided. It tells us something directly about hands (and brains, possibly), but it tells us nothing directly about intelligence. If there is to be anything of substance in evolutionary psychology, it must be able to inform us about the historical evolution of *mind*. But if there is no psychological equivalent of the fossil record, how is this to be done?

Now it has to be admitted that this is not *in principle* an insurmountable difficulty. Indeed, even the way I have set it out is likely to be disputed by Darwinian psychologists and others because it seems to presuppose a radical distinction between the physiological (bones and the like) and the psychological (aggression and the like), and this is a distinction any self-respecting student of the evolution of human beings will deny. And there is good reason to do so. Physiology, behaviour, feeling and culture are all intertwined, and it is their interconnection that allows us to hypothesize from the non-psychological (physical fossils) to the psychological (states of mind), and from the psychological to the cultural (artefacts and codes of conduct, for instance).

To agree that we can hypothesize from the physiological to the psychological to the cultural, however, is not to agree that we can do so with much confidence. Whatever may be true in principle may nonetheless be lacking in practice, and indeed it is here, as it seems to me, that evolutionary psychology is at its weakest.

For example, consider Pinker's attempt at 'reverse engineering the emotions' (Ch. 6). A number of important elements come into play in the story he proposes. First, there is an account of the role of emotion in the life of the mind. 'The emotions are mechanisms that set the brain's highest level goals. Once triggered by a propitious moment, an emotion triggers the cascade of subgoals and sub-subgoals that we call thinking and acting' (p. 373). Second, there is an account of 'our recent ancestors', the nomadic hunter-gatherers, and the environment in which their (and our) psychology was formed.

> Homo sapiens is adapted to two habitats. One is the African savanna, in which most of our evolution took place. For an omnivore like our ancestors, the savanna is a hospitable place compared with other ecosystems. . . . Our ancestors, after evolving on the African savanna, wandered into almost every nook and cranny of the planet. Some were pioneers who left the savanna and then other areas in turn. . . . Others were refugees in search of safety.
>
> (Pinker 1997: 375)

(Wilson subscribes to the same view. 'Early human beings . . . filled a special ecological niche; they were the carnivorous primates of the African plains. They retained this position throughout the Ice Age as they spread into Europe, Asia, and finally into Australia and the New World' (Wilson 1995: 91).)

Putting these two elements together produces a third:

> the key to why we have emotions. An animal cannot pursue all its goals at once. If an animal is both hungry and

> thirsty, it should not stand halfway between a berry bush and a lake. . . . The animal must commit its body to one goal at a time, and the goals have to be matched with the best moments for achieving them.
>
> (Pinker 1997: 373)

The reverse engineering theory of emotion, then, is this. Among nomadic hunter-gatherers, living in the African savanna a long time ago, those with genes that inclined them to have emotions such as fear and disgust (the two Pinker discusses at some length) were at an advantage from the point of view of survival, because emotions, by setting goals for action, prevent risky indecision or even fatal paralysis in the face of competitors and predators. These genetically advantaged nomads were our genetic ancestors, and that is why we have emotions and why our emotions operate as they do.

Thus baldly, but I think accurately, stated, it is evident that the reverse engineering theory of emotion is, to say the least, ambitious. To be maintained with any plausibility it needs to call upon a psychological theory of emotion that is both well formulated and supported by contemporaneous evidence of 'how the mind works'. It then needs to come up with evidence of the mentality of our ancestors in the African savanna. It further needs to show that the postulated mentality of that time satisfied the 'basic survival mechanism' I claimed to be essential to the Darwinian framework of explanation even on the most charitable interpretation. After that, the reverse engineering theory has to be able to show that the 'modern' mentality is a continuation of the 'nomadic' mentality that preceded it. Only then will the claims of evolutionary psychology with respect to (some of) the emotions be made good.

Now it might reasonably be asked how Pinker (or anyone else) knows all this. As Holmes Rolston remarks, 'Most of what the earliest humans thought is lost in the mists of the past; any psychoprehistory is speculative' (Rolston 1999: 292). On what evidence, then, is such speculation based? The answer appears to be that the scope of the evidence drawn upon is very wide and varied. At one level, this is precisely what is to be expected. As we have seen already, Wilson's first excursion in sociobiology led him to expand his inquiry into anthropology and cultural studies, and this is an expansion he now regards as a forerunner of the consilience of all the sciences in a grand collaborative endeavour (*Consilience* being the title of another influential book). Evolutionary psychology, too, is essentially multidisciplinary, emerging as it did from a combination of forces between the anthropologist John Tooby and the psychologist Leda Cosmides. Now Pinker himself is an experimental psychologist. One might have supposed, therefore, that his contribution to the assemblage of relevant materials would consist in experimental results. In fact, somewhat surprisingly, many of his claims about how the mind works turn out upon inspection to be assertions. Perhaps, as he implies at several places, these are assertions based upon experimental evidence (though he also thinks that 'even the most recondite scientific reasoning is an assembly of down-home mental metaphors'; p. 359). Let us suppose that it is so. Could we make such evidence available for the second of the strands I identified – the nomadic mentality of which, allegedly, the modern mind is the inheritor?

On this point E. O. Wilson is clear, if not entirely confident. Acknowledging that 'the archeological evidence from two million years ago' is inadequate, he maintains that 'instead, we must rely on data from the living hunter-gatherer societies, which in their economies and population structure are closest to the ancestral human beings' (Wilson 1995: 133). And in accordance with this belief he cites a great deal of anthropological evidence about a great many tribes – the Munduracu headhunters of Brazil, the

Western Shoshoni, the Yanomamö of southern Venezuela, the Owens Valley Paiute, the Great Basin tribes, etc., etc.

For the moment, and for the sake of the argument, let us assume that contemporary hunter-gatherer societies provide good evidence for how hunter-gatherers operated long ago. (It is a point to be returned to shortly.) Even if this gives us some psychological continuity between the African savanna and the contemporary world, it does so only for some small parts of it. How are we to sustain a psychological continuity not only with contemporary hunter-gatherers, but with the denizens of modern Western societies? On this point Pinker appeals to evidence derived from psychological experiments.

> In experiments on human habitat preference, American children and adults are shown slides of landscapes and asked how much they would like to visit or live in them. The children prefer savannas, even though they have never been to one. The adults like the savannas, too, but they like the deciduous and coniferous forests – which resemble much of the habitable United States – just as much. No one likes the deserts and the rainforests. One interpretation is that children are revealing our species' default habitat preference, and the adults supplement it with the land with which they have grown familiar.
>
> (Pinker 1997: 376)

Lest we falsely imagine that some species of nature mysticism is being invoked here, Pinker is quick to reassure us. 'Of course, people do not have a mystical longing for ancient homelands. They are merely pleased by the landscape features that savannas tend to have.' The evolutionary psychologist's point, rather, is that this preference is biologically 'built-in'. The same conclusion, it seems, is borne out by another piece of empirical evidence. A survey of 'the professional wisdom of gardeners, photographers and painters' revealed, apparently, that 'the landscapes thought to be the loveliest are dead-ringers for an optimal savanna' (*ibid.*, p. 376).

It is hard not to be somewhat sceptical about all this. While it may be naturally tempting to think that contemporary hunter-gatherers are just like their counterparts of forty thousand years ago, there is in fact no reason to make this identification. To suppose otherwise is to imagine that, with respect to such people, time has stood still, in some sense. Interestingly, this is precisely the (now much despised) view that nineteenth-century explorers and missionaries had: that the peoples they encountered were 'primitive' and had failed to make any advance upon the stone age in the way in which 'civilized' peoples had. Of course, the fact that this once fashionable view is now deeply unfashionable is not in itself a reason to reject it. More compelling is the fact that time does *not* stand still, in any sense whatsoever, and accordingly there simply is no reason to think, as Wilson does, that the remaining hunter-gatherers of today provide us with some sort of 'window' through which to view the pre-historical past. The mentality and social behaviour of twenty-first-century hunter-gatherers *may* be identical with that of early human beings living on the African savanna; we have no way of knowing that it is not. But by the same token, we have no way of knowing that it is.

Actually, there is some empirical evidence that must add to doubts about the whole conception. To begin with, there is reason to doubt the unique position of Africa in the evolution of *homo sapiens*, and some evidence that human beings as we now know them emerged independently in Australasia. Second, while sociobiology and evolutionary psychology generally assume that hunter-gatherers are in some sense the most primitive form of human society, and that it is out of and upon such groups that more complex and sophisticated societies have arisen (agrarian to mercantile to industrial), this is far from clear. The anthropologist

Tim Ingold points out that in certain respects, hunter-gatherer societies are more complex and sophisticated than agrarian ones.

> On the basis of a comparative survey of the toolkits of hunter-gatherers, farmers and herdsmen, Oswalt was able to refute the common assumption that hunters and gatherers have simpler tools than any other human groups. In fact the most complex tools were found among specialized hunters, especially hunters – like the Inuit – of large aquatic animals, who have to use considerable ingenuity to obtain inaccessible and potentially dangerous prey. The herdsman, who has ready access to comparatively docile animals faces nothing like the same technical challenges, and his toolkit is correspondingly simpler.
>
> (Ingold 2000: 366–7. The reference is to Oswalt 1976)

Ingold goes on to observe that even this comparison of tool complexity can be misleading, since a different kind of complexity can also enter the subject. In short, the sequence 'simple, less simple, complex, more complex' upon which the sociobiologist and evolutionary psychologist rely is not easily established. The point to underscore for present purposes, though, is that Wilson, and evolutionary psychology more generally, gives a special place to human beings at a certain stage of development – the denizens of the African savanna – and takes the anthropological evidence of certain contemporary groups – hunter-gatherers – to be directly relevant evidence. But not only are we lacking any reason to make such an identification, a proper understanding of the anthropological evidence leaves it quite uncertain that the contentions of the evolutionary psychologist would be borne out in any case.

What of the experimental evidence Pinker cites? Here, if anything, the evolutionary psychologist's position is even weaker. To derive anything at all from it, we must not only accept that the evidence about American children's choices of landscape is as reported, but be willing to generalize from this extremely limited sample to the whole of the human race over a very long period. This is a huge methodological assumption, of course, and remains so even if very many more experiments of the same sort were conducted with the same result. But even having made it, we are not much further forward. Why should we suppose that the explanation for the preference that these children exhibit lies deep in their genetically determined psychological makeup? There is no reason to think so. It could as easily be the outcome of social acculturation. Pinker seems to think that a contrast between the preferences of children and those of adults (in the passage quoted) is telling here. But it is not. Compared with adults, children prefer stories about fairies, magic and so on, all of which they have not seen. But it is at least plausible that this is the outcome, not of a psychological inheritance not yet corrupted by personal experience, but of a cultural differentiation between literature for children and literature for adults.

But even if we discount these doubts about the empirical evidence upon which sociobiology/evolutionary psychology is based, there are two further points of considerable difficulty in the way of this approach. The first is this. Does the 'genetic hypothesis' actually explain *mentality*, as opposed to mere *behaviour*? As we have seen, the basic model holds that specific mental traits, for the most part peculiar to human beings, are to be accounted for by the postulation of a genetic predisposition to such traits. The genes persist and the corresponding traits develop because of the advantage they bestow in the struggle for survival on those who have them. As a general sketch this is not implausible, but how *precisely* is this explanation of the mental supposed to work? Consider the following example. A creature that is able to recognize and reason from evidence (the footprints of potential predators or potential prey, say) and who is liable to corresponding emotional states (fear and excitement, presumably) is more likely to escape the predator or to catch the prey than a creature without these aptitudes. So at any rate the story goes. But why does the evolutionary advantage require realization in the form of mental representations, chains of reasoning or felt emotions? Surely a creature who simply reacted instinctively to the presence of footprints, not with mediating thoughts but with immediate action, would enjoy precisely the same advantages? Why is mind, properly so called, needed at all?

Now this question signals a huge topic of discussion and debate which it is not possible to enter into here, namely, the naturalization of mind. Is it possible to explain the existence and function of mind in wholly naturalistic terms, that is, terms drawn from natural science (whether biological or physical)? Since we cannot take up this issue here without expanding the book to three or four times its intended length, the following remarks will have to suffice. Sociobiology and evolutionary psychology are committed to the in principle possibility of the naturalization of mind. In this, of course, they are at one with a great many other scientists and philosophers, and indeed with the general cast of the modern mind. However, as my brief illustration is meant to indicate, naturalistic explanations of mental phenomena are easily sketched but hard to detail convincingly. In short, the ambitions of the scientific naturalism that Wilson for one expressly espouses, considerably exceed its accomplishments. Evolutionary psychology assumes that wholly naturalistic explanations of mental phenomena such as consciousness, thought and emotion are possible, and its endeavours have been devoted to supplying them. So far, no widely accepted explanations have been devised, and sooner or later this must raise a question about the assumption upon which their search is based; is scientific naturalism adequate to these purposes? It is a topic to be returned to in Chapter 4.

The second difficulty is that a crucial factor – the acquisition of culture – appears to be left out of the account altogether. It seems evident that in the struggle to survive, some advantageous characteristics are acquired rather than inherited. In short, phenotypes learn. Zoological observation confirms that this is true of very many creatures, and that new generations are systematically inducted into ways of nest-building, hunting and so on, that are directly related to their survival. Moreover, studies of song-birds and primates, for instance, give us reason to think that such techniques are embedded in a wider context of behaviour that it is not fanciful to call 'culture'. But if this is true of birds and apes, the same point applies indefinitely many times over to human beings, whose early life consists in an induction into a linguistic, social and cultural inheritance of techniques and technologies that are directly, and manifestly, related to the business of survival and procreation. The chances of modern Western human beings surviving and procreating are vastly greater than those of human beings of any other time or place. This is a result of modern technology in food production, medicine and public health, however, not the result of a different genetic base. On the contrary, the recent ability to write the human genome shows that in genetic terms there is virtually no difference between the healthy long livers of the American Midwest and the relatively fleeting lives of city slum-dwellers in India.

If this is true (and it seems not only incontestable but uncontested), it follows at once that culture presents genetic explanation with an important limit, not just with respect to the explanation of culture itself, but with respect to the survival of living things and the propagation of their genes, because, in the modern world especially, these, importantly, are as much the outcome of cultural as of genetic factors. It should be added, however, that the conception of evolutionary psychology that this might lead us to reject is only one among a possible number (see Caporael 2001). Caporael suggests that there is an important 'continuum between what evolutionary psychology is and what it could become' (*ibid.*, p. 622). Perhaps so, but for the moment we have to conclude that the lines of thought made popular and prominent by Pinker, Wilson and others are seriously limited.

In the face of such a limit, however, some would contend that while genetic explanation may be limited in this regard, evolutionary explanation is not, and that in an important sense culture can be 'Darwinized'. This is our next topic. But before that we should take stock.

Taking stock

Darwinism is committed to the view that the life world in all its manifestations is to be explained as the result of gradual changes over very long periods in the course of which random genetic mutations are perpetuated and developed by natural selection in accordance with the advantages they bestow in the struggle for existence upon the creatures that possess them. In its modern version, the ultimate survivors are the genes themselves, but genes can only survive in animal and vegetable embodiments.

What the argument so far has revealed is that the battle with respect to some of these claims is over. That species have evolved over a very long period of time seems incontestable, and correspondingly, the assertions of creationists to the contrary are not only implausible, but hopelessly hollow to anyone who has a half-decent familiarity with modern biology. But it is erroneous to infer from this, as many Darwinians have tended to, that the evolutionary principle of explanation, even aided by the additional resources of game theory, population studies, anthropology and experimental psychology, can carry all before it. While it is true that some standard objections to Darwinian explanation – the definition of 'fittest', the occurrence of 'altruistic' behaviour, the continued existence of sterile beings and so on – do not present the sort of difficulty they are commonly thought to, there are nonetheless major challenges which Darwinism cannot obviously overcome. The first of these is physical and relates to the claims of gradualism. There appear to be *some* 'irreducibly complex' biochemical mechanisms which at present cannot easily be accounted for in terms of the slow accumulation of gradual changes, and most tellingly, perhaps, this seems to be true of the origins of life itself.

To concede that this is so is not necessarily to take the next step that its best-known advocate – Michael J. Behe – takes. We can allow that there is irreducible complexity without thereby supposing that we have evidence of intelligent design in nature along the lines proposed by pre-Darwinians such as Paley. (It is this further step that has brought down upon Behe the anti-creationist wrath of the Darwinians.) Intelligent design apart, it remains the case that if and insofar as irreducible complexity, whatever its ultimate explanation, is a *fact*, it constitutes a major obstacle to the ambitions of 'universal Darwinism'.

Another no less significant challenge is to be found at the level of psychology. It is important not to overplay dichotomies here. There is every reason to regard the physical, psychological and social natures of human beings as importantly intertwined, but this acknowledgement falls far short of the ambitious contention that patterns of human sensation, thought and feeling are the outcome of genetic structures acquired in the course of evolution many thousands of years ago. In the first place, it is not at all obvious just how the relation between the perpetuation of genes, the nature of the mental and the accumulation of advantages in the struggle for survival is supposed to work. In the second place, neither sociobiology (to some of whose contentions there will be reason to return) nor evolutionary psychology is in a position to adduce any relevant evidence in favour of the intriguing suggestions they advance. In any case there is a further gulf to be bridged. How could we plausibly deny that when it comes to human behaviour, cultural inheritance is at least as great an influence as genetic inheritance?

Memetics

Actually, not all Darwinians do deny this. On the contrary, some positively assert it. In *The Selfish Gene*, Dawkins remarks:

> As an enthusiastic Darwinian, I have been dissatisfied with the explanations that my fellow-enthusiasts have offered for human behaviour. They have tried to look for 'biological advantages' in various attributes of human civilization. For instance, tribal religion has been seen as a mechanism for solidifying group identity, valuable for a pack-hunting species whose individuals rely on co-operation to catch large and fast prey. . . . These ideas are plausible as far as they go, but I find that they do not begin to square up to the formidable challenge of explaining culture, cultural evolution and the immense differences between human cultures around the world. . . . The argument I shall advance, surprising as it may seem coming from the author of earlier chapters, is that, for an understanding of the evolution of modern man, we must begin by *throwing out the gene* as the sole basis of our ideas on evolution. I am an enthusiastic Darwinian, but I think Darwinism is too big a theory to be confined to the narrow context of the gene. The gene will enter my thesis as an analogy, nothing more.
>
> (Dawkins 1976: 191; my emphasis)

This is indeed a surprising turn. It amounts to the concession that 'the selfish gene' is not the key to human existence after all. At best, what we learn from genetics, applied to culture, provides us with illuminating analogies. What are these analogies with? In answer to this question, Dawkins invents a concept (and a term) – memes – that has been explored at greater length by others.

> What after all is so special about genes? The answer is that they are replicators. The laws of physics are supposed to be true all over the accessible universe. Are there any principles of biology that are likely to have a similar universal validity? . . . Obviously I do not know but, if I had to bet, I would put money on one fundamental principle. This is the law that all life evolves by the differential survival of replicating entities. The gene, the DNA molecule, happens to be the replicating entity that prevails on our own planet. There may be others. . . . But do we have to go to distant worlds to find other kinds of replicator and other, consequent, kinds of evolution? I think that a new kind of replicator has recently emerged on this very planet. . . . It is still in its infancy, still drifting clumsily about in its primeval soup, but already it is achieving evolutionary change at a rate that leaves the old gene panting far behind.
>
> The new soup is the soup of human culture. We need a name for the new replicator, a noun that conveys the idea of a unit of cultural transmission, or a unit of *imitation*. 'Mimeme' comes from a suitable Greek root. . . . I hope my classicist friends will forgive me if I abbreviate mimeme to *meme*.
>
> (Dawkins 1976: 191–2; emphasis in original)

The idea of the meme – a cultural analogue to the gene – has been taken up with enthusiasm in some quarters, and regarded with considerable scepticism in others. Stephen J. Gould is prominent among the sceptics. The parapsychologist Susan Blakemore, author of *The Meme Machine*, is the most ardent of the enthusiasts. Dawkins himself, it is worth noting, has expressed substantial reservations in subsequent writing (see Dawkins 1982: 112). What is at issue is this question: Could there be a science of memetics comparable to the science of genetics? One of the first people to examine this question directly and at length was the philosopher Daniel Dennett (in *Darwin's Dangerous Idea*). Dennett thinks that

> There is no denying that there is cultural evolution, in the Darwin neutral sense that cultures change over time, accumulating and losing features, while also maintaining features from earlier ages. . . . But whether such evolution is weakly or strongly analogous to, or parallel to, genetic evolution, the process that Darwinian theory explains so well, is an open question. . . . At one extreme . . . it could turn out that cultural evolution recapitulates *all* the features of genetic evolution. . . . At the other extreme, cultural evolution could be discovered to operate according to entirely different principles. . . . In between the extremes lie the likely and valuable prospects: that there is a large (or largish) and important (or merely mildly interesting) transfer of concepts from biology to the human sciences.
>
> (Dennett 1996: 345–6; emphasis in original)

Accordingly he concludes that

> The prospects for elaborating a rigorous science of memetics are doubtful, but the concept provides a valuable perspective from which to investigate the complex

> relationship between cultural and genetic heritage. In particular, it is the shaping of our minds by memes that gives us the autonomy to transcend our selfish genes.
>
> (Dennett 1996: 369)

On the face of it, Dennett's middle ground between the two extremes sounds like a sensible position to adopt on the 'open question' of memetics, and his idea of 'transcending the selfish gene' is a topic worth returning to. But the sweet reasonableness of this *via media* crucially depends upon our being able to make sense of Dawkins's original suggestion. Can we do so? What is a meme, and to what is it analogous? A partial answer to the first of these questions is to be obtained by noting the kinds of things that believers in memes have in mind. Dawkins offers us the following examples – 'tunes, ideas, catch-phrases, clothes fashions, ways of making pots or of building arches' (Dawkins 1976: 192). Dennett's list is longer and more ambitious. It includes such particulars as the alphabet, Mozart's *Marriage of Figaro*, chess, the theory of evolution and 'Greensleeves' (Dennett 1996: 344), but also what he calls 'very general memes' such as 'music, writing, calendars, education, environmental awareness, arms reduction . . . anti-Semitism . . . computer viruses' (p. 363). One aspect of such lists is striking – their sheer variety. Can it really be true that the theory of evolution and the tune 'Greensleeves' are, in any sense whatever, the same kind of thing? On the face of it, this seems wholly implausible, but perhaps more importantly, the variety raises doubts about the concept. Memes, it appears, are discrete cultural units that replicate by imitation and persist by surviving a process of selection. Now it seems quite plausible to regard a computer virus as a discrete cultural artefact that replicates itself, but not so plausible that it does so by imitation; a second copy of a virus does not come into existence by imitating the first. On the other hand, while it seems plausible that groups describing themselves as 'environmentally aware' might indeed imitate the behaviour of other groups who think of themselves in this way, in so far as they promote different specific causes, there is no obvious element of replication; a public campaign against road building may be very different from a campaign against whale hunting. And when it comes to calendars and alphabets, we have to ask which calendars and alphabets. Is there any reason to group together the modern Western alphabet with the 6000 characters regularly used in Chinese writing?

In my view, these are not just quibbles. If memetics is to go anywhere, we need to know, quite literally, what we are talking about. And a brief survey of the literature suggests that we do not. Even if we did, of course, there is still the second question – to what are memes analogous? According to Robert Aunger, the editor of *Darwinizing Culture* (significantly subtitled 'the status of memetics as a science'), 'we currently have at least two rival paradigms contending for dominance in memetics – the "meme-as-germ" and "meme-as-gene" schools'. At a formal level – in epidemiology and population genetics – there may be some common ground between the two, but as explanatory analogues the implications of each are quite different. Genes 'leap' from phenotype to phenotype, according to Dawkins, but they do so by means of reproduction. The 'leap' takes place by one existing phenotype giving rise to another not yet existing phenotype. Germs, by contrast, 'leap' from one existing phenotype to another already existing phenotype. Do we inherit memes, or do we 'catch' them?

Actually, in the original passage from Dawkins there is a clue. 'Cultural transmission', he remarks, 'is not unique to man. The best non-human example I know . . . [is] the song of a bird called the saddleback which lives on islands off New Zealand' (Dawkins 1976: 189). The songs of these birds, apparently, can be classified in 'dialect' groups, and each young bird derives its own peculiar repertoire not from genetic inheritance but from hearing and imitating a selection acquired from the 'collection' of songs available. Furthermore, every so often, this imitation is imperfect, and through repeated imitations the imperfection is magnified to the point where a recognizably 'new' song emerges, to be added to the common stock upon which new generations of song birds can draw. Something similar to the birds' behaviour happens among human beings. We learn to sing tunes from a common stock, and the more 'catchy' tunes win out in a sort of 'natural selection' because they stick in the mind better than those that are not so catchy. Such tunes can 'leap' from person to person in the sense that if I start humming a tune, now or later you too may involuntarily start humming it. Sometimes, of course, the imitation by others of even common catchy tunes is imperfect, and from this fact new tunes (can) emerge.

The involuntary humming of tunes that appear, unbidden, in the heads of those who hum them is a familiar phenomenon, and Dennett also makes extended use of the 'tune in the head' paradigm by recounting this kind of experience, and then re-describing it as the spread of 'a horrible musical virus, at least as robust in my meme pool as any melody I actually esteem' (Dennett 1996: 347). Now as it seems to me, this example serves very well to reveal the basic thought behind memetics – that all cultural transmission might operate in this way. But if this *is* the basic thought, the proper response to it is plain: all cultural transmission does *not* operate in this way, and a great deal of it *could* not. The 'musical virus' case is one that fits the (germ-school) memetic picture, not perfectly, but tolerably well. Very few of the other phenomena on either Dawkins's or Dennett's list fit the same picture. How, precisely, are we to specify 'environmental awareness', 'arms reduction' or 'education' such that one manifestation of the thing in question can clearly be said to be the replication of another? And even where there is a distinct unit of a sort followed by another unit of the same sort, it is not at all obvious that the relation between the two is one of replication by means of imitation. Two performances of the *Marriage of Figaro* are clearly two instances of the same thing. But why should I regard the second as a replicating imitation of the first? It is at least as plausible (though not in my view correct) to say that both are replications of the original score. And with popular tunes there is this further difficulty: Is a sophisticated orchestral arrangement of 'Greensleeves' no less a replication of the meme than my involuntary humming?

What memetics leaves out above all, of course, is the *deliberate* recovery and manufacture of cultural forms and artefacts. Just to stick with the musical case: contemporary musicians of the 'authenticity' school have gone to great trouble to recreate both the instruments and the music of sackbuts and crumhorns which, in the process of musical selection, had long ago been eliminated. Others, of course, have striven for compositional forms whose merits aim to lie precisely in their being unprecedented. And the same point could be made about almost any potential 'meme'. It may be the case that there are distinguishable cultural units that replicate themselves by means of imitation and survive and evolve in a process somewhat analogous to that of natural selection. For my own part I see no special reason to deny the possibility of such a thing. But I see no necessity to assert it either, and it seems as plain as anything can be that any attempt to fit the origins, development, survival and decline of all that is properly described as culture into this pattern is destined to fail, and that the determination to do so is nothing less than perverse.

Unlike Blakemore and the other wide-eyed enthusiasts of the 'science of memetics', neither Dawkins nor Dennett hopes to do this. Dennett, as we have seen, thinks that the talk of memes is interesting chiefly to the extent that it enables us to think about a mental world beyond genetics. Dawkins inclines in the opposite direction and holds that the chief interest in the idea of memes probably lies in the additional light it might throw on genetics. In either case, of course, we are licensed to leave the topic of memetics, either to pursue the very much wider issues that will be the subject matter of Chapter 4, or to continue our investigation into the significance of the genetic revolution, which is what the next chapter aims to do.

3

GENETIC ENGINEERING

To a considerable extent the conclusions of the previous chapter were negative; the intellectual difficulties facing neo-Darwinism are formidable, and the explanatory promises of universal evolution almost certainly false. Such conclusions might give rise to the impression that genetic biology is something of a dead duck. Nothing could be further from the truth. If Copernicus initiated an irreversible scientific revolution by replacing a geocentric view of the world with a heliocentric one, Darwin effected no less great a revolution with *The Origin of Species* and *The Descent of Man*. This is a revolution that has been strengthened immeasurably by the later alliance with genetics. Whatever the difficulties in neo-Darwinism, and however unjustified its claims to universal and unlimited explanatory power may be, the fact remains that there can be no return to pre-Darwinian ways of thinking. A certain sort of creationism has been intellectually defeated *for ever*, and nothing is to be gained for ethics or religion in attempting to resurrect it.

But in any case, the importance of genetic biology is not restricted to the explanations it may or may not lead us to. Modern genetics has practical as well as theoretical significance, and in fact to many minds, this is where its greatest importance lies – in what it enables us to *do* rather than what it enables us to explain.

The genetic revolution, in other words, is technological as well as scientific. It has revolutionized not only the ways in which we seek to *understand* the world (by enabling us to explain phenomena that had hitherto been a mystery), but also our attempts to *change* the world (by engineering outcomes barely conceivable hitherto). Moreover some of these possible outcomes are so dramatic that there is reason to wonder whether the genetic revolution will not make a reality of what Aldous Huxley could only imagine – a 'brave new world'.

Huxley's novel of that title, of course, was more a warning than a promise, and it is this dichotomy that has structured almost all the discussion surrounding genetic engineering. Is the future that genetics makes possible, a nightmare or a dream? This is the guiding question of the present chapter.

By and large those who regard a genetically driven future with enthusiasm are inclined first of all to draw attention to the advantages of genetic screening, to point to the diseases that we might detect at conception, and the defective births – spina bifida and other gross malformations – that we might thus be enabled to prevent. It is a small step from this, of course, to the suggestion that, once genetic defects have been identified, the same technologies might allow not merely for warning but for remedy, and with this step we enter the realms of genetic modification. But why should genetic modification stop at the remedying of defects? Why not, if we can, attempt positive improvement? With this further step, genetic modification becomes genetic design, and it is at this juncture (for many people) that dream turns to nightmare, and the Pandora's box of 'Frankenstein foods', human clones and 'designer babies' is opened.

One obvious way of examining the issues surrounding genetic engineering is to consider just this progression – from genetic screening through genetic modification to genetic design. And by considering each of the steps in turn we can try to determine whether there really is a crucial transformation, and if so, at what point it occurs, and what makes it crucial.

Genetic screening

> Consider a gene on chromosome 18 that suppresses colon cancer . . . a tumour suppressor whose location has not quite been determined for sure. It was thought to be a gene called *DCC*, but we now know that *DCC* guides the growth of nerves in the spinal column and has nothing to do with tumour suppression. The tumour-suppressor gene is close to *DCC*, but it is still elusive. If you are born with an already faulty version of this gene you have a much increased risk of cancer. Could a future genetic engineer take it out, like a faulty spark plug from a car, and replace it? The answer, quite soon, will be yes.
>
> (Ridley 1999: 244)

This passage from Matt Ridley's popular and successful book *Genome* captures a common picture of how things stand in hi-tech biology. Now it is obvious that 'a future genetic engineer [could] take it out, like a faulty spark plug' only if it could first be detected. Though the term 'genetic screening' was originally applied to the business of testing populations for the presence of a particular genetic defect, increasingly it is used to mean a genetic 'health check' administered to individuals, one that they might choose or be compelled to undergo, and whose purpose is precisely to spot 'faulty' genes. It is with this second sense of 'genetic screening' that we are concerned here because of the ethical problems that its deployment is thought to raise.

It might be argued that the position is already a little more advanced than Ridley describes, since some of the outstanding uncertainties he alludes to – the precise location of the gene in question for instance – are closer to being resolved than they were when he wrote, thanks to the completion of the human genome project in 2000. For several decades microbiologists have been accumulating information about the genomes of hundreds of different kinds of entity, from bacteria to plants, insects and animals. Despite a widespread impression to the contrary, at present the precise sequence is known for only a very few bacteria and plants, and now for a single primate – ourselves – which is why the completion of the *human* genome was an occasion of such special acclaim.

The precise implication of this project's having been completed is a matter of considerable importance to the present discussion. There is an unspoken implication behind Ridley's example; if we can identify the gene *DCC* as the one 'that guides the growth of nerves in the spinal column' and identify another 'tumour-suppressor gene' that is 'close to *DCC*', then, armed with our new knowledge of the whole human genome we can make indefinitely many such identifications. In short, Ridley (along with many others) assumes that what can successfully be done in some important instances – the identification of a gene with a specific manifestation – can be generalized such that for every ailment there is a rogue gene, and for every physiological function there is a genetic base. This may not be so. One of the surprises that resulted from the completion of the genome project related to the *number* of genes which, as the project progressed, declined dramatically. At the outset it was believed that the human genome would comprise several hundred thousand genes. As the project neared its end 100,000 was thought to be more likely. When the project was complete, the number was reckoned (provisionally) to be something over 30,000, though this may be revised upwards.

Now such a number seems impossibly small if we are to identify a gene for each and every feature and function, as well as every defect, that goes into the makeup of human beings. More striking yet is this fact. The number of genes in the human genome is not so very much larger than the number in that of much less complicated beings. And when it comes to the primates, there is very little genetic difference indeed – a 1 per cent difference in DNA between humans and chimpanzees for instance (though variation in DNA is not the same as variation in the number of genes, it should be noted). The picture is complicated by the fact that our knowledge of many of the genomes between which such comparisons are made is largely based upon estimate and conjecture. As I noted earlier, the full sequence is actually known for very few biological entities. There is the further complication that the number of codings within a gene can vary enormously. While it is the case that the number of genes between one species and another may be close, the complexity of each gene can be very different. This fact allows for there to be explanations of variety that are still genetic, though within rather than between the respective genomes.

Nevertheless it does seem to be the case that numerical genetic variation is surprisingly small. Some people have concluded from this that (to adapt the words of the Psalmist) we are little higher than the animals, but this is a mistaken inference. The differences between humans and other animals are, and remain, what they have always been. People are not more like apes than they were before we discovered the genetic commonality of the two. Whatever geneticists discover, the physiology and behaviour of the humble tapeworm is as distant as ever it was from our own.

What follows from the discovery that humans are not so genetically different from other animals is that the explanation of the huge gap between humans and all other living things is unlikely to be found at a genetic level. For instance, it seems that the brains of humans and those of apes start out at much the same size. The difference lies in the capacity of the former for far greater growth. No doubt this potentiality has a genetic foundation; the apes don't have the kind of brains that can grow that big. But the importance of the difference does not lie in the resultant *size* of the organ, but in the range and scope of what the adult with such a brain *knows*. What makes the difference between adult human and adult ape is not brain size, but the vast disparity in what each has learned. The same point applies with respect to differences *between* human beings. 'The' human genome is, of course, a generalization, because human beings vary considerably in stature, colour, size and behaviour. But it is possible to write a single, universal human genome because, though there are obvious and manifest differences between, say, Africans and Japanese, at a genetic level the differences are minuscule. This discovery was widely heralded as confirmation of the ethical idea of the 'brotherhood of man', that 'underneath' so to speak, we are all the same and should thus be treated equally. But once again, however desirable the ideal, this is a mistaken inference. The differences between African and Japanese remain as they were. All that follows from the discovery that the human genome is virtually the same in all human beings is that the explanation of very many of these differences is unlikely to be just genetic.

Of course, as we saw in the last chapter, even the most ardent advocate of 'the selfish gene' – Richard Dawkins – concedes this, hence the introduction of the concept of 'meme' and the investigations of sociobiology and evolutionary psychology. But it is a concession that is inclined to be forgotten when we move from explanation to engineering. Biotechnology has put into our hands a dramatic power to engage in genetic manipulation, and to do so in pursuit of longstanding human purposes such as the promotion of health. This fact alone, however, tells us nothing about the scope of this new power. The sort of case that Ridley imagines may become a real possibility. The point to emphasize, though, is that there may not be so very many such cases.

Still, important though it is to acknowledge this, we should focus on the potentiality that there is. What possible techniques could we employ for the almost fantastic genetic cutting and pasting that Ridley's example invokes?

> [F]ortunately nature [has] already invented them for her own purposes. The glue is an enzyme called ligase, which stitches together loose sentences of DNA whenever it comes across them. The scissors, called restriction enzymes, were discovered in bacteria in 1968. . . . In 1972, Paul Berg of Stanford University used restriction enzymes in a test tube to chop two bits of viral DNA in half, then used ligases to stick them together again in new combination. He thus produced the first man-made 'recombinant' DNA. Humanity could now do what retroviruses had long been doing: insert a gene into a chromosome. Within a year, the first genetically engineered bacterium existed: a gut bacterium infected with a gene taken out of a toad.
>
> (Ridley 1999: 244–5)

It would be hard not to be impressed by this astonishing biotechnology, and indeed by the applications that have already been made of it, most notably, of course, in the production of genetically modified organisms of a relatively simple sort (a context we will return to). However, in assessing its significance we must be careful not to confuse the technically possible with the practically possible more broadly conceived. The picture Ridley paints is one in which techniques for dealing with genetic illnesses such as cancer that were formerly (at most) distant aspirations have recently become real possibilities. But to turn these possibilities into actual therapies we have to possess more than technology; we have to have social systems of health care. It is not so much in the techniques of genetic screening, cutting and pasting that the real difficulties lie, but with their embodiment in the wider context of health care.

The first of these difficulties is this. The real availability of technical remedies is everywhere a function of resource allocation, and for those who do not have the necessary resources, no remedy is available. This point is not to be confused with a plea for distributive justice and a fairer world. Rather, be the system as fair as you will, the absence of resource can render the technically possible practically valueless. Consider a familiar example. One of the most striking recent technical advances in medicine is that of organ transplantation. It is important to observe, however, that recourse to organ transplants as a remedy is severely limited not only by the normal financial constraints that apply in any real circumstance, but by the limited number of organs available for transplant. Imagine a world in which the techniques of organ transplantation have been invented and successfully mastered, but where, as a matter of fact, no organs are actually available. In such a world, organ transplantation would be a *technically* possible but not a *practically* possible solution to failing hearts, kidneys, livers and so on. From this we can infer that the technical is not always practicable, and hence that the technical and the practical cannot be identified.

So too with the case that Ridley imagines. Genetic screening and consequent genetic 'cutting and pasting' as a way of preventing colon cancer in later life may indeed be a technical possibility. And who could deny that the prevention of colon cancer is a good thing? Further, who, in consequence, could deny that the advent of such a technology is to be welcomed? And who, as a result, could hesitate to advance its cause? So, at any rate, an ineluctable chain of reasoning seems to go. But in this, as in many ethical issues, we are morally obliged to engage in a measure of cost/benefit analysis, and in particular to take account of opportunity costs. Resources devoted to developing, exploiting and extending the sort of screening/treatment Ridley envisages are resources *not* devoted to other beneficial therapies. Ethical objections to genetic screening as a general practice are most usually raised either with respect to the activity in itself or to its immediate consequences. But the proposal that this should be the *general* practice raises additional questions about the desirability of incorporating genetic screening into health care services. In reality, the cost of providing such screening and treatment for every foetus even in wealthy Western countries would be enormous, and made prohibitive (it is plausible to think) when we consider how many more diseases we might try to screen for. Since there are indefinitely many fearful diseases, it is desirable to screen foetuses for all of them. The task in prospect is (at present) nothing short of Herculean. New technologies may change this, though it is worth observing that if experience is anything to go by, new technologies *increase* the number of possible procedures and do not diminish them. The chief point, however, is that the possibility of indefinitely many legitimate calls upon screening is something that has to be weighed into the ethical calculation.

At a minimum we can say that before we could rightly welcome the sort of scenario Ridley describes writ large, we would have to engage in a highly complex calculation, one which involves not only the issue of opportunity costs, but also that of the probability of the benefits. In reality, of course, we are not talking of eliminating all possibility of the disease, but of reducing the chances of its developing. This alters the calculation importantly. Weighing the prospect of colon cancer against its absence is the easy case. But what if it is a matter of *reducing the chances* of colon cancer (or any other life threatening condition) rather than actually preventing the disease? How do we set the value of merely reduced risk of harm (as opposed to harm itself) against the real cost of alternative benefits forgone? Of course, these are not questions peculiar to the issue of genetic screening; they arise with respect to a great many health care strategies and related public policies. Indeed an entire subject – health economics – has grown out of the need to address these crucial issues. The point to stress here is that it is only in the light of certain answers to questions of this kind that genetic analysis and correction would be transformed from technical to practical possibilities for most people in the real world. In other words, if we want to think about the rights and wrongs of genetic screening, we have to think about it, in large part, in a social and economic dimension.

Now it could be argued that, though sound enough, this sort of point can hardly get to the ethical heart of the matter. What we want to know first and foremost (many readers will suppose) is whether the thing is right or wrong in itself and in prospect, and not whether it proves too costly retrospectively. It is worth observing, however, that even if the cost of genetic screening restricted its deployment to a very small number of people, this would still introduce an important new element to the consideration of its merits. Once we pass beyond a single case of screening/correction and think of it as a medical therapy, then however limited its availability and use may be, the calculation of probabilities and the weighing of outcomes will have to play a part in the assessment of its value. A close parallel is to be found in the following example. It is already standard practice in many countries to offer pregnant women amniocentesis to determine whether the foetus has Downs syndrome. (It is an example we will return to in Chapter 4.) If the test proves positive, the woman is offered a termination. But conducting the test itself has the effect of increasing the chance of miscarriage. The result, according to one calculation (which some dispute), is that the number of additional normal foetuses lost in miscarriage as a result of the test is about four times that of abnormal foetuses whose lives are surgically terminated. In other words amniocentesis is not itself causally neutral with respect to relevant outcomes.

If this parallel is a good one, then we should expect that the responsible assessment of the merits of genetic testing requires us to abandon the deceptively simple scenario Ridley presents in terms of the single case – colon cancer and what, technically, can be done about it. In other words, there is not nearly as much ethical interest in asking whether a one-off act of genetic screening and replacement is right or wrong in and of itself, as there is in asking whether we should employ genetic screening as a standard medical therapy. Responsibly assessing a general practice of screening is a complex matter involving empirical facts, estimates of probability and trade-offs between different kinds of harms and benefits. The inclination of many people – journalists especially, but scientists and ethicists as well – is to focus on the sort of single case that Ridley describes, and thus to arrive at a principled 'yes or no' answer with respect to the question of its being right or wrong. But once we grasp the importance of understanding genetic screening to be a putative *practice* rather than a possible *action*, it becomes clear that the moral issue is inseparable from wider issues of social policy and health economics.

In short: it may be tempting to confine the discussion of genetic screening to the one case/one disease scenario, but to do so is to distort the real ethical issues which are much wider, less novel, more complex and less susceptible to clear-cut answers than people who take sides on these issues would want. It is true, however, that there is at least one intrinsic aspect of gene therapy of the kind envisaged that raises a special difficulty. It is evident that the interconnectedness of the genome must be of the first importance. That is to say, the range of phenomena to be explained and the limited number of genes in terms of which to explain them, implies that there is a high level of interaction. This means that even if we can identify a gene peculiarly connected with colon cancer or bone brittleness (remembering always that its presence is only probabilistically connected with the actual occurrence of the disease) we cannot be sure what the other consequences of removing it might be. And we could not really discover this without performing the 'cutting and pasting' and waiting 30 years or so to find out. To take this approach, however, is to experiment on human lives, in a way that hardly seems defensible. Who are to be our guinea pigs? By the nature of the case they would have to be human beings at a stage of development long before that at which they could give informed consent. Furthermore, if the moral justification for Ridley-type procedures lies in the estimated benefit that they are expected to bring to the lives of those on whom they are performed, this justification is seriously eroded, if not eliminated, by the possibility of the inestimable harm they might also do. In short, it is only if we isolate the procedure in terms of expected beneficial consequences, and ignore the potential for harmful ones that we can arrive at a positive conclusion. And, of course, this begs the question: we can hardly be justified in undertaking an action for the sake of its outcomes if we only consider the beneficial ones. Whether we like it or not, potential harms must also figure in the calculation.

What are we to conclude from this? It is always a mistake to become morally enthusiastic about or exercised by fantasies. This is the danger that attends any general discussion of genetic therapy. The plain, if disappointing, truth is that general discussion based upon technical possibilities has little moral mileage in it. The real issues lie with specific proposals in concrete circumstances, and the proper examination of these issues will have to deal with matters of fact and probability that cannot be ascertained in the abstract.

Genetic information

There is another widely discussed issue that arises with respect to genetic screening. This relates to what we might call 'the ethics of information'. Whatever its therapeutic benefits may or may not be, genetic screening brings to light information about individuals that we could not obtain in other ways. Ethical questions arise with respect to both the right to know such information and the uses to which such information is put. These are matters to be examined separately.

Suppose screening shows up some genetic defect that is likely to have very serious consequences for the individual concerned. Should the person be told? A common response to this question is that people have a right to this information – that they should be told, *if they want to be*, and that it is an unacceptable exercise in medical paternalism for doctors (and others) to keep this important information from the patient. That may be so. But I shall not go over this familiar ground again because a more interesting question, in my view, is this: *Should* they want to know?

It is tempting to answer this question positively by invoking a general epistemological principle that self-knowledge is better than ignorance. Now no doubt in many contexts this is a plausible principle to invoke. But is it true in general? And does it apply to genetic self-knowledge? Why *should* I give preference to knowledge over other values? Sometimes affirmative answers implicitly assume that knowing one's condition is a necessary precondition to amending it. In many of the cases we are concerned with, however, this need not be true, for the very good reason that there *is* nothing to be done. Once more Ridley's example is misleading. Perhaps the identification and replacement of a gene which causes cancer of the colon is relatively easy; I am not well enough informed to deny this, though I know that others do. But it is important to stress that the 'simple fix' case (if this is indeed one) is very rare. The ability to test for sickle cell anaemia, for instance, was established long ago, and yet, decades later, the existence of the test has been followed by nothing in the way of a remedy. The same is true of Huntington's chorea. We can detect those who are likely to develop this disease. But having detected them there is nothing we can do to alleviate, still less prevent, their dreadful fate.

This second example bears closer examination. Huntington's chorea, everyone knows, brings about a ghastly death in early middle age. Why should I want to know that this is my fate? In reality, since it has long been known that the disease runs in families, I may at present know that there is a good chance of my developing it. What reason could I have for wanting to employ the new screening techniques which genetics has made possible to turn this possibility into a certainty? There can be such reasons, of course, relating to child bearing perhaps. But these are case specific; they do not warrant a principle that commends knowledge over ignorance in general. 'Ignorance is bliss', a common saying runs. At any rate, genetic ignorance may save a lot of anguish.

There is a further complication. Taking a test and remaining in ignorance of the result may induce as much anxiety as knowing an unfavourable outcome. If so, I have reason not to take the test. There is nothing irrational in preferring lower levels of anxiety to higher levels of knowledge. A reasonable person could easily arrive at this preference. Now it is worth noting that this and the former point can be made about any sort of predictive test and are not restricted to genetic screening. From this we may conclude, as it seems to me, that as far as the choices that face us are concerned, the advent of hi-tech genetic testing has introduced nothing new to a recurrent aspect of moral and practical deliberation in the human condition. Knowledge and anxiety, health and happiness, hope and fear attend our every endeavour and have to be balanced and traded off against each other in the constant business of deciding what it is best to do. Genetic screening and genetic manipulation are new. But the problems and choices they feed into are not. So far as the first-person reception of genetic information is concerned, then, no new ethical issues confront us. Life can be difficult. But we knew that. Genetics has not made it intrinsically more difficult.

There is also the third-person point of view, of course. This returns us to the question of the uses to which genetic information is put. A recurrent fear in the genetic age is that information will be gathered, or demanded, that could be used to deny medical insurance to those who are found to have certain genetically based conditions. It is not difficult to articulate the idea behind this fear. If it is possible to determine whether a given individual will develop a disease such as cancer or Alzheimer's in later life, insurance companies would have evident reason to deny him or her insurance cover with respect to those conditions. Moreover, since this is obviously information relevant to its actuarial calculations, such a company would also have reason to require genetic testing before agreeing to issue a policy. John Harris believes that such an outcome is inevitable.

> We should be clear from the outset that it will not be possible to keep the new and ever increasing data generated by genetic screening out of the hands of insurance companies for the simple and sufficient reason that they can demand it as a condition of contract. If it is withheld, insurers will simply deny insurance cover, and of course

> false or incomplete information will vitiate the policy and automatically render the candidate uninsured.
>
> (Harris 1998: 261)

This prediction cannot be quite as certain as Harris alleges, because the fear that genetic information will be misused is so widespread that some states have already taken steps to put legal limits on the information which insurance companies can require. But what is this fear exactly? Generally speaking, it is the fear that some people will be unable to insure their lives and livelihoods, not because of a life-style choice (smoking, promiscuous sex, etc.) that they might choose to give up, but because of a genetic inheritance about which they can do nothing, and that this is unfair.

Now a great deal turns on two crucial points. Harris argues that increasingly people will fall into one of two groups, 'the uninsurable' and 'good risks', and that 'good risks will qualify for . . . reduced premiums . . . at the expense of those excluded from insurance altogether or those with heavily "loaded" premiums' (Harris 1998: 262). He continues:

> The existence of these two groups will of course make life much easier for insurance companies. The *risk* which hitherto has been of the nature of insurance business will progressively be reduced. We have already witnessed a substantial shift from *assessment of risk* as the basis of insurance to *minimization of risk*. This is the process whereby insurance companies, instead of accepting that the name of the game is to assess the risk of a particular policy and set premiums accordingly, have gradually moved to attempts to minimize or even eradicate, rather than simply assess risk. This process will inevitably accelerate with the ideal being the

> dismissal of risk altogether, insurance becoming a form of regular saving.
>
> If (or perhaps when) this process is completed, we will have moved from *insurance against all risks* to *insurance only of the risk free*.
>
> (Harris 1998: 262–3; emphasis in original)

Harris is here predicting the future with a confidence that seems to me unwarranted. Consider his final contention only. If there is no risk to the insured, there is no incentive to take out insurance policies, with the result that there is no profit in offering insurance of this sort. In such circumstances we may expect companies to change tack. And there are indeed other possible scenarios. An insurer has reason to charge very high health insurance premiums to someone known to have a high probability of Huntington's chorea (for which there is no cure) if the cost of care in the final stages of the disease is very high. But there would be reason to offer the victims of this disease very attractive terms on pensions, which, the reasoning goes, they are unlikely to collect. In fact, there would also be reason to *reduce* the cost of health insurance since, compared with the potential health care claims of the longer lived, the amount to be paid out on victims of Huntington's is likely to be small. The two sums might balance out. In any case, the prognostication is only probabilistic. Some people so identified will not develop the disease, and/or in the interval some effective treatment for Huntington's may be discovered. In either event the insurer is the loser and the individual the gainer because cheaper pension funds are drawn upon for far longer than the actuaries expected.

Huntington's chorea is a very special case, one of only a few where clear genetic screening is possible and the outcome highly predictable. But even in these special cases, the possibility of survival into old age is something that insurance companies must include in their calculations. Indeed, we have some evidence to go on here, because just such calculations lie behind the different insurance and pension schemes on offer to smokers and non-smokers. Since the evidence connecting tobacco smoking with lung cancer is very good indeed, Harris's scenario would lead us to expect that smokers would by now have been relegated to the class of the uninsurable. This is not so. On the contrary, many companies offer smokers *better* pension schemes on the (some would say cynical) grounds that they will die early. And this parallel further alerts us to the fact that insurers, like everyone else, must take into account the 'dialogue' between physiology and life style. All tobacco smokers have a much higher risk of lung cancer, but not all develop it. Similarly, not all those who have a genetic disposition to heart disease, colon cancer or Alzheimer's disease will develop it. However high the probability, something depends on diet and other factors. To repeat a familiar but important adage, nothing in this life is certain. Huge advances have been made in genetic understanding, but they have not altered this basic truth, a truth that properly informed actuaries will incorporate into their calculations. The result is unlikely to be as detrimental to policy holders as has been anticipated. The group identified by Harris as 'uninsurable' may be so for a time, but many of the people who comprise it will have characteristics that different actuarial calculations may well show to be profitably insurable. If so, any legal constraints that have been put in place to prevent the dissemination of genetic information may be not only premature, but detrimental.

This more promising scenario is not any more 'inevitable' than Harris's, of course, though I believe it to be likely. But however the future may turn out there is this further question: why is the use of genetic information by insurance companies so objectionable? Harris has an answer: 'It is clearly grossly inequitable to deny people the benefits of insurance because they are prudent enough to be screened and monitored for health risks' (Harris 1998: 263). This is an assertion, not an argument, and prefacing it with 'clearly' merely underlines the assertion. *Why* is it grossly inequitable? If we regard the insurer as a free agent, then he/she must be free to enter or not to enter a contractual arrangement. Of course, let us agree that unfair discrimination, which is to say discrimination on irrelevant grounds such as skin colour or ethnic origin, is as objectionable here as anywhere. But the grounds of discrimination in this case – susceptibility to illness, chances of mortality, etc. – are not irrelevant. An unspoken assumption in Harris's approach to this question of insurance, one which is widely shared I think, is that there is an unfair transfer of burden from the relatively sickly, higher premium payer to the relatively healthier lower premium payer. This picture needs to be made more complex, however. If the relatively healthy pay less into the system, they also take less out. To force the insurance company to offer more equitable premiums, accordingly, is to introduce inequalities at the level of benefits. In any case, we cannot assume that the relatively healthy necessarily have higher incomes. When they do not, such enforced equality of insurance premiums would result in the poor subsidizing the rich.

In general, as it seems to me, the assumption is made that the availability of genetic information will work against those who are identified as having special susceptibilities. It is an assumption that might be thought to be implicit in Phillip Kitcher's phrase 'The New Pariahs', the title of one of the chapters in his book about the impact of 'the genetic revolution' (see Kitcher 1996). Kitcher himself, however, is alive to the possibility of quite different scenarios, and consequently adds a question mark to his chapter title – 'The New Pariahs?'. One of the distinctions he draws is illuminating here.

Genetic tests might disclose two quite different kinds of facts

> about applicants: their risks of acquiring various debilitating diseases, whether or not they work in the environment the employer provides (general health risks), and susceptibilities to diseases that might be triggered by substances emitted in the workplace (workplace-specific risks).
>
> (Kitcher 1996: 145)

Now while it could be the case that insurers refuse to insure people with certain general health risks, it could also be the case that they offer much more favourable terms to individuals and companies in cases in which genetic testing has been used to identify work-specific risks and eliminate the relevantly damaging substances in the workplace. While the first clearly puts the uninsurable at a disadvantage, the second is equally clearly in their interests. Yet genetic screening is a necessary pre-condition of both, from which we should conclude, *contra* Harris, that nothing is 'inevitable' here. Just how things work out and whether there are any 'new pariahs' or not, is a matter for empirical discovery, not *a priori* speculation.

It is true, of course, that the different ways in which the widespread use of genetic information works out are likely to lead to different distributions of social burdens and benefits. Furthermore, though the prospect of widespread genetic testing is usually greeted with moral anxiety, it can reasonably be asked whether genetic screening might not be an obligatory instrument in ensuring a fairer distribution of these benefits and burdens. Here we touch upon a topic of great interest, but also one of considerable complexity – namely the impact of genetic technology in general on questions of distributive justice. However, the space its proper treatment would require means that I shall not pursue it further here. Readers who are specially interested will find a very comprehensive discussion of the issues in *From Chance to Choice: Genetics and Justice* by Allen Buchanan and others. The points to be made here are two. First, the scenario that Harris and others anticipate is certainly not a foregone conclusion. The availability of genetic information about individuals in the contexts of insurance and employment may promise as much as it threatens. There are scenarios in which the outcome is far more favourable to those who are found to have genetic potential for degenerative and/or fatal diseases. Second, the question of the 'grossly inequitable' cannot simply be declared or assumed to favour one side – those to whom cover is denied. There may be something grossly inequitable about *forbidding* insurance companies to seek genetic information if it has the result, for example, that resources are in effect transferred from the poor to the rich.

This 'problem' about genetic screening and insurance has been discussed quite widely (a further discussion will be found in Suzuki and Knudtson: 1989, Ch. 7), but there is another interesting issue raised by genetic screening that has gone almost unremarked. This relates to the possibility of genetic 'ethnic cleansing'. In the ordinary way, any attempt at ethnic cleansing is morally objectionable because of the violence and death it implies and involves. In the Serbian conflict of the late 1990s, for example, the (futile) attempt at ethnic cleansing was an exercise in brutality whose wrongness was both evident and incontestable. But suppose it was to be accomplished not by the destruction of existing people, but by the prevention of their being born.

In reality, since there is no biologically plausible version of anything called 'race' – there simply are no human 'races' – genetically engineered ethnic cleansing is not an issue, whatever the irrational (if powerful) hatreds of one group of people against another. But there are other possibilities of a similar sort – 'sexual cleansing' for example. In 1993 a considerable stir was caused when Dean H. Hamer and a team of researchers from the US National Cancer Institute announced in *Science* that 'We have now produced evidence that one form of male homosexuality is preferentially transmitted through the maternal side and is genetically linked to chromosome region Xq28' (quoted in Peters 1997: 95). This claim was widely represented as the discovery of the gene for homosexuality. In fact it is more modest than this, and there is no consensus yet around even the more limited claim, still less the ambitious one upon which the press seized. However, for present purposes let us suppose that homosexuality is genetically based, and that in principle it is possible to 'extract' the homosexual gene and replace it with a heterosexual one. For the prospect of sexual cleansing to come into view, we have only to add to this technical possibility certain, easily imagined, political circumstances. Suppose that in some country or other the following is democratically endorsed as public policy: pregnancy must be standardly accompanied by a procedure that detects and then 'treats' potential homosexuality.

There is good reason to believe that such a policy would in fact be fruitless. Such a 'homosexual' gene, if there is one, would show itself only in the homozygote and not in the heterozygote. But the heterozygote will still be carrying the gene in a recessive form. Since 50 per cent of the population are heterozygotes, this means that there is no real prospect of such 'sexual cleansing'. But what, if anything, would be wrong with making the attempt? The question raises several of the issues that surround the idea of 'playing God' that will be the subject of the next chapter. The attempt at sexual cleansing (and this is only one of several possible examples), however futile it might actually be, carries the implication that someone – the state, the medical profession, 'the people' – is in a position to determine what populations are 'fit' for existence. But equally important is the assumption of biological determinism. To attempt to create a 'brave new world' by means of the genetic manipulation of populations implies that such manipulation guarantees the desired outcome. But if, as I argued, universal Darwinism rests upon some fundamental errors, there are and will always be patterns of human behaviour that are *not* subject to genetic manipulation. My own view is that homosexuality is one of them, though it would take more space than is here available to argue this case in particular. The point may be generalized beyond this particular example, however. It is often said that with the advent of genetics a whole host of new moral dilemmas and difficulties faces us. Just what moral problems really do confront us is not a matter to be settled independently of the seemingly more 'theoretical' question of the scope and power of genetic explanation. The fear, and the hope, is that we have finally arrived at a position in which we can fashion the world in accordance with our preferences. This crucially depends upon the prior question of how far we have really understood the nature of the world we propose to re-fashion. In short, the topics of this chapter, though often considered in relative isolation, are inextricably linked with the topics of the last. What it makes sense to try to engineer genetically depends upon what we can successfully explain genetically.

Genetic modification

The example with which the last section closed – the attempt to eliminate homosexuality – leads naturally to the next range of topics because it imagines a circumstance in which we use genetic knowledge not merely to remedy defects but to engineer positive genetic outcomes (or, as in the foregoing example, those that are perceived to be so). The high point of such ambitions is 'designer babies'. This is a topic to which we will return, but it is better to start with the more modest subject of genetically modified organisms – GMOs – one of whose most widely discussed applications, of course, has been in the production of 'improved' vegetables and animals for consumption.

It is worth emphasizing at the outset that genetic engineering of this kind is not, in itself, new. Something of the sort has been going on for thousands of years – from the time agriculturalists first selected certain seeds and plants rather than others for the purposes of propagation. Nor is modern biotechnology novel in its 'scientific' approach to genetic engineering, if by 'scientific' we mean a systematic investigative approach. The eighteenth century, chiefly in Britain, saw a sustained and highly self-conscientious effort to improve plants and animals for agricultural and horticultural purposes by selective breeding. Indeed, it is something of an irony that the cluster of (not always related) ideas and aspirations that make up modern environmentalism often include a passionate opposition to GMOs and an equally passionate defence of historic breeds. In reality many of these breeds were the outcome of deliberate genetic engineering. Of course, the animal breeders of the eighteenth century and the pigeon fanciers of the nineteenth knew nothing of the science of genetics. That is to say, they knew how to engineer certain outcomes, but they had no understanding of the biological processes that underlay their success. The emergence of genetics revealed these, and it also put into the hands of human beings the power of engineering genetic outcomes with greater speed and precision. Does this make any ethical difference?

In the abstract it is difficult to see that it could. Although GMOs are widely represented, and regarded, as singular novelties of post-genetic biology, the simple truth is that they have been with us a very long time. The spectre of 'Frankenstein foods', upon which media interest has focussed so much attention, should rightly have been drawn to the public's attention several centuries ago. And even the methods are not new. Cloning, which is to say the creation of new generations that replicate their adult forebears and are not produced through natural reproduction, is a positively ancient practice. Cloning plants (under the less scientific names 'layering' and 'taking cuttings' or 'slips') has been a familiar part of horticulture for hundreds, perhaps thousands of years. It has been claimed, for instance, that the celebrated wine grape Cabernet Sauvignon was developed by selective cloning as far back as the time of Julius Caesar. Now if there was no special reason to worry about genetically engineered variety produced by cloning then, why do we have occasion to worry now?

This question provides us with a useful argumentative strategy. If we assume that the 'scientific' husbandry that marked agriculture and horticulture in the eighteenth and nineteenth centuries (and much earlier in the case of plants) is unobjectionable, we can explore possible objections to modern biotechnology by asking whether there is any relevant difference between the two, and if so, where it lies. This is, of course, a parallel with a venerable precedent. The very first chapter of Darwin's *The Origin of Species* is devoted to the subject of 'Variation under Domestication' and the term 'natural' selection takes its meaning from the contrast with 'deliberate' selection by pigeon fanciers and nurserymen. Indeed, it is precisely by employing this parallel that Darwin most clearly articulates his central thesis.

> [W]e see in man's productions the action of what may be called the principle of divergence, causing differences, at first barely appreciable, steadily to increase, and the breeds to diverge in character both from each other and from their common parent.
>
> But how, it may be asked, can any analogous principle apply in nature? I believe it can and does apply most efficiently, from the simple circumstance that the more diversified the descendants from any one species become in structure, constitution, and habits, by so much will they be better enabled to seize on many and widely diversified

> places in the polity of nature, and so be enabled to increase in numbers.
>
> (Darwin 1996: 93)

Similarly we might ask: if 'man's productions' by the methods of selective breeding are not to be decried, how can a change to biotechnological methods make a difference? I shall consider four possibilities – that modern biotechnology has undesirable objectives, that it involves an unacceptable 'tinkering' with nature, that it is environmentally uncontrollable, and that it requires a kind of research that is ethically suspect. I shall then return to examine the assumption on which this strategy rests and ask whether the older forms of genetic engineering are indeed unobjectionable.

In considering the first line of objection it is to be observed that the earlier form of genetic engineering, no less than the modern one, was firmly focussed on the better satisfaction of human desires. Improved beef cattle, dairy herds, varieties of wheat and barley all claimed attention because of their ability to serve human purposes – animals with a higher proportion of usable meat, milk or wool, crops with higher yields that would grow more effectively in certain conditions. Nor were these attempts at genetic engineering confined to promoting human welfare narrowly conceived. Fanciful ideas such as inserting the gene of a flounder to produce a 'square tomato' that is more convenient to pack and easier to slice (something that is widely, but falsely, supposed to have been done) are sometimes represented as flippant departures from serious purpose. Yet horticulturalists have long been engaged in the development of new varieties of plants whose purpose is to enhance the aesthetic appearance of ornamental gardens rather than the nutritional resources of the kitchen. Horses and dogs were selectively bred for working purposes, but they were also bred for riding and hunting, and

Darwin himself was specially interested in the breeding of racing pigeons, whose ultimate rationale lies in the advancement of sporting pleasure. 'Believing that it is always best to study some special group', he tells us, 'I have, after deliberation, taken up domestic pigeons' (Darwin 1996: 19).

The observation that the selective breeders of the past had purposes beyond those of producing better food and clothing significantly undermines the suggestion that modern biotechnology is chiefly objectionable for the ends it pursues. The quest for the square tomato is neither more nor less trivial than the quest for the black tulip. One striking difference between the two is speed. Another is precision. By the old methods it took several generations to produce a variety, and hence quite a lot of time to secure a specific outcome – creamier milk or whatever. Moreover, the whole process was subject to a high degree of uncertainty in its results. In the past, anyone who sought to produce a better strain of sheep, say, was subject to a good deal of trial and error and had to wait some considerable time before the effort paid off, if it did. Modern laboratory methods, by contrast, are both more precise and more efficient. This difference can be exaggerated. The experiments in question are extremely sophisticated and can only be carried out in a few laboratories world-wide. Nor is the process so very efficient; producing the cloned sheep Dolly took 277 attempts. And its speed is somewhat limited by the fact that the progeny of such experiments can develop to maturity no faster than those generated naturally. Still, despite these qualifications, there are significant potential efficiency gains, or else the experiments would not be undertaken. But why should this matter, morally speaking? If the aim is the same (and acceptable) how could the time it takes to accomplish it constitute an objection? Indeed, the opposite seems to be true – that the increase in speed and efficiency should be welcomed, if the end is a good one. We can appeal to a plausible abstract principle here, that if a thing is worth doing it is worth doing more efficiently.

The fact, of course, is that the greater efficiency of modern biotechnology is not universally welcomed. While the United States has generally accepted extensive genetic modification of vegetable foods, popular (and political) resistance to GMOs is widespread in Europe and sustains a large part of the 'green' movement. This connection brings us to the second possible objection – that biotechnology involves an unacceptable 'tinkering' with nature or the natural order.

A striking instance of such tinkering is the creation of special types of animals for medical and other research purposes. Among the most common and least controversial are modified bacteria. (No one seems to worry very much about these.) One of most startling (to ordinary ears) is the embryo of a headless frog. But though in some ways more repellent, this perhaps is not as morally controversial as the (patented) Oncomouse, the mouse genetically modified so that the mechanisms that prevent the development of cancers (in all of us) do not work in the mouse. The idea, of course, is to aid research efforts in trying to find a cure for the disease in humans. But can it be right deliberately to bring into existence a creature that contains the seeds of its own demise? Or more accurately, is it right to bring about the existence of a creature deprived of some of its most important survival mechanisms?

What *exactly* is wrong with developing the Oncomouse? Here is one line of thought worth considering. Such a creature is not merely unnatural in the sense of not being found in nature. There has been built into it, so to speak, the characteristic that its development run counter to a fundamental principle. This is the principle that the biology of a thing should be so ordered as to promote and maximize its flourishing. To take steps that deliberately deflect it from this goal is a form of perversion. A parallel might be this. Injustice is largely a matter of causing injury and violating rights, and justice a matter of reparation and vindication. But wrong also attaches to 'perverting the course of justice' even where the action so described does not itself cause injury or violate a right. Now the unfolding of the Oncomouse's biology does seem to be approximately as follows. We use our knowledge to bring it about that things do not go as they should, that the process malfunctions. Cancer is the *going wrong* of cells; that is what makes a cancerous growth more than just a protuberance like a nose or an arm, say. Accordingly the Oncomouse has been modified *in order that the tendency to go wrong* will not be checked, and it cannot be right deliberately to engineer the wrongness of things. In short, this little creature is a natural monstrosity of a kind, and that is where the evil of its intentional creation lies.

On first acquaintance at any rate, this is quite a powerful argument in my view. If it is to be elaborated effectively, however, the underlying idea of natural goods and evils needs to be spelt out at greater length and in a wider context. It is a topic I shall reserve for the next chapter. For the moment, and continuing to pursue the argumentative strategy of this section, we will confine ourselves to asking whether the Oncomouse (or any similar creature) is radically different from the less dramatic productions of the older form of genetic engineering. And I think that the answer is that it is not. Consider the case of the seedless grape (which may have been produced by the old form of cloning). Such a strain (as opposed to a one-off freak) for obvious reasons could never be found in nature. If 'flourishing' is a natural end of biological structures so too is propagation. Consequently, deliberately to produce a seedless grape is to create something that (in the absence of human intervention) is bound to fail in its natural function. In short, it is an aberration consciously contrived. A similar point might be made about the mule – a sterile cross between the horse and the donkey brought into existence as a 'biological tractor', so to speak.

Of course, there would appear to be an important difference – seedless grapes don't suffer. Whether there truly is this difference is an empirical matter. That is to say, whether a cancerous mouse suffers is not something we can determine *a priori*, and not something we can simply assume. I can imagine ways in which an inherently cancerous creature developed for research purposes might not have to suffer. Perhaps it is sufficient for the tumours to be at a pre-pain stage, at which time the mouse is killed painlessly. (Those who know more about these things than I do tell me that this is indeed generally the case.) Suffering, in other words, is something we have to find out about from case to case.

But let us suppose that the Oncomouse does suffer. This takes the case into the realms of normal moral deliberation. Is this *necessary* suffering? Does it serve an end good enough to outweigh the evil of the suffering? Are people more important than mice? All these are good and interesting questions, but they arise in many different contexts and have nothing special to do with modern genetic technology. The embryonic headless frog, the Oncomouse, and many other extraordinary creations of the biological laboratory (the fly with eye-covered legs is among the strangest) look and sound strikingly new. At one level they are; before the advent of modern genetics such creatures could not have been created. This is their novelty. But in itself this does not imply any real *moral* novelty. Moral novelty arises only when morally new dimensions come into play. Now the argument about natural goods and evils that I rehearsed briefly does imply such a dimension – our responsibility as procreators – a hugely interesting and important subject closely connected to popular fears that genetic engineers are (in danger of) 'playing God' (the theme of the next chapter). But as far as the present discussion goes, the issues central to it – the comparison between old and new engineering methods, the causing of pain, and the balance of animal suffering over medical benefit – involve nothing that we could call moral novelty, and must take their place among the somewhat imprecise semi-empirical arguments about relative cost and benefit that surround other animal welfare issues. These are real choices, certainly, but they are not *Unprecedented Choices* (contrary to the title of a recent book on these themes).

To summarize: We began the comparison of new methods with old in order to explore four possible lines of objection – that modern biotechnology has undesirable objectives, that it involves an unacceptable 'tinkering' with nature, that it is environmentally uncontrollable, and that the research it requires is immoral. With respect to the first we can say that modern laboratory methods serve much the same aims as those of the selective breeders of previous centuries; that is to say, they seek improved varieties of plants and animals for the betterment of human (and sometimes other animal) life. Moreover, these varieties are sought in pursuit of human purposes of different kinds, some to do with health and nutrition, others with art and sport, some very substantial, some rather trivial. With respect to the second claim – that genetic engineering is 'tinkering' with nature (in an objectionable sense) – we have seen that while some research laboratory animals can plausibly be described as 'freaks', at one level of analysis they are no more so than the mule and the seedless grape. At another level, they may indeed represent a failure to observe what I have called 'our procreative responsibilities'. But if so, this can be sustained only after the careful consideration of complex ideas (postponed). This leaves the third and fourth lines of objection, which will now be considered in turn.

Environmentalism

In his 1998 Amnesty Lecture on 'Cloning People', the philosopher Hilary Putnam remarks that

> [E]ven techniques of 'bioengineering' that seem utterly benign, such as the techniques that have so spectacularly increased the yields of certain crops, have the side effect, when used as widely as they are now, of drastically reducing the genetic diversity of our food grains, and thus increasing the probability of a disaster of global proportions should a disease strike these 'high yield' crops.
>
> (in Burley 1999: 3)

Putnam here expresses one version of a common anxiety – that genetic modification of plants and animals is wrong, not so much in itself as for the impact it has across the environment in which it takes place. This impact is usually thought of in two ways. The first is that to which Putnam alludes – the diminution in genetic diversity. The second is the creation of uncontrollable organisms. Consider the second first. This is unquestionably a danger, and it is one that is at the forefront of most environmentally inspired anti-GMO protests. The idea is easily enough stated. Artificially manipulated and modified organisms cannot be confined to single fields or farms. The wind, the weather, birds and insects can all play a part in spreading these organisms beyond the area they were intended to occupy, and hence beyond the human mechanisms intended to control them. Once out in the open, so to speak, their impact upon other organisms is highly unpredictable.

The spread of GMOs into the environment is commonly referred to as 'pollution'. To describe it in this way begs several questions, not least the basic question of whether the world into which they 'escape' is in any real sense a realm of purity, and hence a realm prey to pollution. But, however this may be, as a matter of fact, such present evidence as we have suggests that the risk of GMOs 'escaping' is not so very great. Moreover, steps can be taken (and are now regularly taken) to ensure that the modified organisms do not thrive except under certain highly specific conditions. The technology is so sophisticated, in fact, that it can incorporate environmental safety devices – seeds that fail to germinate, plants that die except in specially contrived circumstances. The most obvious of these devices is built-in sterility (though this in itself may raise other concerns, chiefly that of giving a monopoly to the suppliers of improved grain). Moreover, if we are to examine the issue fully, there are environmental benefits as well as risks to be considered. Although 'organic' is often represented as an alternative to 'genetically modified', genetically modified crops can reduce dependence on the use of chemical herbicides and pesticides. To this degree, if no other, genetic modification can bring 'green' benefits, and is thus to be welcomed rather than feared, from an environmental point of view.

The same point may be made more generally. There has come about an association of ideas that wrongly construes environmental conservation, organic farming and animal welfare as causes properly united in opposition to genetic technology. This association is highly misleading, for the truth is that these concerns can pull apart. An environmental scientist of my acquaintance works with bacteria that have been genetically modified to glow. In the presence of toxins the light diminishes, and as with other biosensors, this property may be useful in the development of a test for detecting poisons in seafood intended for human consumption. The main purpose of this research programme is food safety – the desire to find a more sensitive and reliable test. But a further advantage of its success would be to put an end to the present method of testing which consists in injecting large numbers of mice with liquidized shellfish and noting their convulsions until paralysis causes respiratory failure and they die. Toxic levels are correlated with the length of time it takes them to do so. Faced with a choice between this test and the one in prospect, one would naturally suppose that concern with food safety and animal welfare should join forces in welcoming genetic modification. Yet for the most part the detractors of genetic modification represent it as the enemy rather than the ally of food safety and animal welfare, and this is simply not the case.

But whatever the actual impact and potential of genetic modification with respect to welfare, and whatever the risks and dangers attaching to GMOs, it is of considerable importance to note that exactly the same environmental risks and dangers have arisen from much less scientifically advanced methods in horticulture and agriculture and at a time when genetic modification in the laboratory was still some considerable time off. In his book *Why Things Bite Back*, Edward Tenner has documented in an absorbing way the environmental disasters that have regularly been induced by schemes of agricultural improvement. Just one such instance is that of kudzu.

> Kudzu (*Pueraria lobata*), a semi-wild leguminous vine native to Asia, thrives in the most neglected and abused soils. . . . During the New Deal the U.S. Soil Conservation Service advocated kudzu as a plant that could restore Southern cotton lands devastated by insects, erosion and the Depression. They were right to admire its vigor. Like weeds, kudzu flourishes in disturbed and marginal settings, resists insects and drought, needs no preparation of the soil or fertilizer, and above all grows rapidly. . . . Its lush growth keeps the soil cool and moist. . . . Land planted with kudzu loses 99 per cent less soil than land planted with cotton. It restores soil nutrients and makes good pasturage and hay. No wonder the Soil Conservation Service sent 73 million kudzu seedlings to farmers. . .
>
> . . .Yet only ten years later, farmers and federal officials

> alike were condemning kudzu as a nuisance plant. . . . Without diseases or North American insect enemies . . . kudzu has invaded the technological infrastructure of the flourishing eastern Sunbelt with a vengeance. It pulls down telephone poles, blacks out neighborhoods by warming local power transformers until they trip. [S]horts high-voltage lines on long distance electric transmission . . . obliterates traffic signs and spreads over bridges . . . grows over the rails and slickens tracks as trains crush it [and] can overwhelm and envelope nearly every stationary object – unmoved automobiles, side-tracked railroad cars, abandoned houses.
>
> (Tenner 1997: 145–6)

What the example of kudzu shows (and it is only one of many) is that *any* attempt to improve on the natural environment can go badly wrong, with wholly unanticipated and almost unmanageable side effects. Such disasters are not the prerogative of biotechnology and, as Tenner records, predate its advent. There is no doubt to my mind that similar dangers are attendant upon biotechnology, and the solemn assurances of genetic technologists to the contrary are worth no more than the simple-minded advice and assistance of the US Soil Conservation Service in the 1930s. Nevertheless, the general point is that, though biotechnology brings with it environmental risks, on the face of it there is nothing *new* to fear in them.

Many people are inclined to a contrary view on this last point, of course. While they recognize how disastrous episodes such as the kudzu fiasco can be, they also believe that modern genetic technology has much greater potential for disaster. This is because they think that genetic engineering operates at the level of the internal workings of nature and not just its outward manifestation. It follows, or so this way of thinking inclines us to believe, that while kudzu-type experiments don't change things fundamentally, genetic modifications do, because, it is commonly said, they are *irreversible*. Actually, the general point needs some refinement because of the important distinction in contemporary biotechnology (especially applied to human beings) between 'stem cell' and 'germ line' modifications. Whereas stem cell research affects only one generation (the individual for whose benefit it is conducted) germ line changes are replicated in succeeding generations. This is why stem cell engineering meets with notably less resistance than germ line, because it is with germ line modifications that irreversibility seems to come into play (I say 'seems' because some such procedures are reversible). If we deploy stem cells, we are involved only in the well-being of an individual. If we alter the germ line, we are committed to an effect on generations to come.

The precautionary principle

In the articulation of these fears, irreversibility is not the only conception at work. The idea that genetic engineering (of the germ line variety) somehow penetrates below the level to which other types of engineering are confined is also common. Neither idea is easy to articulate with any degree of precision; the introduction of kudzu has proved irreversible, and as we have seen, the genetic engineering of 'whole' domestic plants and animals has gone on for centuries. But whether these ideas can be refined or not, the more interesting aspect of both in the context of concern with environmental risk is the implicit claim that there is a *qualitative* difference to the (germ line) changes which genetic engineering brings about such that the risk attaching to them is not merely one of extensive bad consequences (as in the kudzu case) but of *catastrophe*. This is, I think, what people have in mind when they warn of changes so 'profound' that their possible impact is scarcely calculable. Now from this it is often inferred (or at least implied) that since *catastrophe* is a different order of risk to mere danger, rationality not merely permits but requires a complete ban on any range of activity that is potentially catastrophic, regardless of the actual probability of such a catastrophe. Behind this line of argument lies a rule of reason that has been named 'the precautionary principle'. It is a principle not confined to biotechnology, but one regularly invoked in environmental debates at large, and it is increasingly invoked as the justificatory basis of both national and international laws relating to the environment. Here is a good example of the way in which it is standardly appealed to.

> The IPCC [Intergovernmental Panel on Climate Change] scientists predict, based on their computer models of climate, increases of temperature more than ten times faster than life on Earth has experienced in at least 100,000 years, and probably much longer. . . . But the IPCC scientists may be wrong . . . such are the uncertainties of the climate system that they could, nonetheless – just conceivably – be wrong. . . . But do we want to gamble on that tiny possibility? If the world's climate scientists are in virtual unanimity that unprecedented global warming will occur if we do nothing about greenhouse-gas emissions, would we not best serve our children and theirs if we took heed – even if there are uncertainties? . . . Nobody – repeat, nobody – can deny that there is at the very least a prospect of ecological disaster on the horizon where the greenhouse effect is concerned. Those who choose to ignore the prospect, therefore, wilfully elect to ignore the environmental security of future generations.
>
> (Legget (1990) quoted in Manson 1999: 13)

On first acquaintance, many people are inclined to regard the precautionary principle as little more than common sense. However, as Manson and others have pointed out, it is in fact a highly contentious one because it illicitly supposes that the mere prospect of catastrophe can be weighed into calculations of practical rationality irrespective of probabilities. But this is a demonstrable error, and an error, ironically, that undermines a cautious approach to the environment. The problem is that any given means of averting one catastrophe will have attached to it some 'prospect' of a different catastrophe, with the result that the precautionary principle is either arbitrarily invoked on one side rather than the other, or else prescribes paralysis.

The crucial point is most easily stated abstractly. In order to choose between different possible outcomes we have to make some estimate of their desirability/undesirability and at the same time weigh into our deliberations their respective probability. Now the simplest way of doing this is to multiply the probability of its occurrence by the value (negative or positive) of the anticipated outcome, and choose the larger positive number. (This is the strategy employed by rational gamblers.) The deployment of such a strategy means that we have reason to avoid steps whose negative consequences are very unlikely, provided that the consequences are negative *enough*. The precautionary principle says, in effect, that there are possible outcomes (nuclear winter, global warming, etc.) whose negative value is so great that the merest prospect of their coming about as the result of something we do is enough to give us reason not to do it.

Some writers have noted that there is a formal similarity between this argument for caution and Pascal's celebrated 'Wager' argument for believing in God. According to Pascal, if the potential benefits of believing in God are great enough, then we should believe in Him even if everything else suggests that the probability of God's existence is very small. And of course, if we take the traditional Christian view of the afterlife seriously, it seems that the potential benefits are not merely very great but infinite – an eternally long and immeasurably blessed life. The trick in the Wager lies with the appeal to infinity. It is only infinity multiplied by the tiniest probability that is guaranteed to win the Wager. Unfortunately, however, infinity can also be called upon to point our deliberations in exactly the opposite direction. By becoming believers in God we also run the risk of apostasy (falling into theological error), which we would not do if we remained agnostics, and with it the prospect of hell – i.e. infinite pain and misery. This further consideration, of course, gives us an unsurpassable reason for *not* believing in God; to do so brings with it the risk of the infinitely bad. And this holds good even if we think that the chances of becoming an apostate and being punished by eternal damnation are tiny in comparison with the chances of heaven. In short, the Wager works both ways, and it does so precisely because it trades on the idea of *infinite* benefit, and hence cannot ignore the possibility (however remote) of *infinite* loss.

Now the precautionary principle is in effect making the same appeal to mathematical infinity, even if this is not expressly acknowledged by those who advance it. The argument goes that environmental catastrophe has a negative value of such magnitude that multiplied by even the tiniest probability, it will determine our course of action. But as in the Pascalian Wager, there is no reason to attach the merest prospect of catastrophe to just one action or policy. Consider the example of global warming. Even if global warming were indeed an environmental catastrophe (there are differences of expert opinion about this, it should be noted), and further, even if the production of greenhouse gases could be demonstrated to increase its probability substantially, it still would not follow that the precautionary principle exclusively endorsed severe restrictions on emissions. This is because from time to time climatologists have also wondered if there is not some evidence of the beginnings of a future ice age, which would also be an environmental disaster, to be averted only by the intensification of global warming. If this were the case, then reducing emissions would also be catastrophic and hence ruled out by the precautionary principle. Once more, as in Pascal's Wager, the fact that the probability of a future ice age is far lower than that of global overheating does not affect the argument. 'Infinity' can be brought in as a value on both sides.

Applied to biotechnology the point is equally plain. We cannot rationally advocate a ban on some piece of biotechnology on the grounds that the outcomes could be catastrophic, without at the same time acknowledging that banning it could also be catastrophic, and in this way it becomes evident that any appeal to the precautionary principle results in stalemate. It simply does not determine what we ought to do. Let it be the case that biotechnology tinkers with the basics in a way that earlier technologies have not done. Even so, we cannot conclude that we have reason to resist or reject it. Everything turns on the (no doubt tedious) business of calculating respective costs and benefits. If we are to find some basis for an absolute rejection it must lie elsewhere; the precautionary principle, however much it may appear to accord with 'common sense', does not establish this conclusion.

Genetic research

This brings us to the fourth objection commonly raised against genetic engineering – that it requires unethical research programmes. There are a number of directions from which this point is pressed but I shall be concerned with only two – the use of animals and the use of human (usually embryonic) material.

Compared with times past, the contemporary world, at least in the West, has undergone a significant change in attitudes to other animals. Whereas before the mid-nineteenth century it was accepted without question that animals could be put to any use whatever that was to the advantage or benefit of human beings, the view began to prevail that we have moral obligations to other animals also. It is easy to state one basis for this view. Since morality is called into play by passive suffering as well as by rational action, the important question to ask about animals (as Jeremy Bentham remarked) is not 'Can they reason?' but 'Can they suffer?', and while earlier ages seem to have regarded even the higher animals as unfeeling machines, it does not take much observation to see that animals can indeed suffer. This fact alone does not *settle* any moral questions; it only raises them. But it does call into play the requirement that, if we are to inflict suffering (on humans or animals) we require some justification for doing so. It is from this relatively elementary ground that animal welfare movements sprang up.

In the twentieth century the argument took a new turn when people began to argue (philosophers among them) that animals had rights and that these rights brought moral considerations into play even where human conduct had no obvious impact upon animal welfare. The topic of animal rights, unlike that of animal welfare, is still a highly contentious one. I shall not be concerned further with it here, not so much because it is contentious, but because the contentiousness arises from the complexity of the issues involved, a complexity that requires more space to be given to the topic than is available.

The cause of animal welfare (by contrast) is, in general, uncontentious nowadays. It is widely accepted, I think, that it is wrong to cause unnecessary suffering to animals. The vexed question, of course, is what is to count as 'necessary' and 'unnecessary' in this context. It is not a difficulty that is peculiar to genetic engineering, or research related to it, however. Biological and medical scientists have long experimented on animals for the purposes of human (and veterinary) medicine. The arguments about vivisection (experimental operations on live animals) raged long before genetic experimentation took on the prominence it now has. The new issue, if there is one, we have already touched upon with the example of Oncomouse, where what is at issue is not so much the infliction of suffering as its creation. Now as I noted, it is an empirical matter whether or not such creatures do actually suffer and whether to serve the requisite research purposes they have to do so. We cannot answer this question in general but only in particular cases. Though there are aspects to which there will be reason to return, for the moment this is an important point to stress. As in the case of genetic therapy, there is constant pressure on those who participate in these debates to declare in principle 'for' or 'against' a whole class of actions. But it really is a mistake to do so. There are principles to be invoked and considered, certainly, but particulars also matter, because if our concern truly is with unnecessary suffering, there is an ineliminably factual dimension to the question, and we cannot simply declare the facts to be this way or that. People sometimes suppose that 'it stands to reason' that the Oncomouse and similar creatures suffer. But nothing 'stands to reason' here; everything requires to be investigated.

This, obviously, is not the place to conduct such investigation, since the purpose of this book is to explore a much wider context within which the relevant investigations should take place. However, there is one point about animals and genetic engineering that is worth making. 'Dolly the sheep' won world-wide fame for being the first successful animal clone, an animal artificially produced directly from genetic material with the purpose of duplicating the features of the adult animal from which the material was taken. With the advent of Dolly, cloning became the major topic in debates on the ethics of genetic engineering. Speedily, however, the focus of these debates became the extension of the technique to human beings. We can clone sheep; the cloning of human beings, accordingly, cannot be far away; would it ever be right to attempt it? This is an evidently important question, to be considered further, but it has overshadowed the question of whether it is acceptable to clone animals. And it is worth adding that, in reality, while the cloning of animals, for commercial as well as scientific purposes, is now fairly widespread, the cloning of a developed human being (with predicted dates that regularly recede) has yet to take place. (Late in 2001 an American laboratory claimed to have cloned a human embryo, though others disputed the accuracy of this description.)

The case of ANDi, the first GM monkey (cloned by US scientists in 2001) illustrates the general point very well. One of the scientists involved is reported to have said: 'We don't support any extension or extrapolation of this work to humans. That's not what we're about at all. We're trying to help accelerate the day when innovative cures for disabling and devastating diseases are shown to be safe and effective.' The campaign 'Against Human Genetic Engineering' responded by saying: 'This is yet another step on the slippery slope to designer babies. People should wake up to the fact that genetic engineering of people could be just around the corner. We need a global ban on human genetic engineering' (quoted in the *Daily Telegraph* for 12 January 2001). What is striking about both comments is that ANDi's welfare is not mentioned. Everything turns on the potential impact, for good or ill, on human beings.

Yet if we do stick with animal welfare, there is, as it seems to me, something to be said on the negative side. If we consider more closely the practice of cloning, the result is often the creation of animals with radical defects and deformities, many of which emerge only as the animal grows. In short, suffering creatures are brought into existence. For this created suffering, as for suffering inflicted, morally responsible behaviour requires a justification. The form of this justification is clear – the suffering is outweighed by the benefits. What are these benefits? It is not always appreciated that even yet the actual business of producing laboratory generated animals is very much a hit and miss affair. In the creation of ANDi, 224 eggs were genetically modified, 40 embryos fertilized, five pregnancies induced, with a resulting three live births, only one of which had the property that the scientists sought to give it. And even ANDi was not a complete success from the point of the research programme. The same is true of cloning and if 'hit and miss' is the mark of the method, it is not any the less a mark of the anticipated benefits. The creation of ANDi was remarkable in its way (involving the insertion of a jellyfish gene into a monkey), but there are no identifiable therapies that this might give rise to. Of course it can be said, with truth, that these investigations are at an early stage and that without some such preliminaries no such therapies are likely to emerge. Perhaps so. But it is important to be aware of the dangers of the appeal to ignorance on the part of scientists no less than proponents of 'the precautionary principle'. From the fact that 'one never can tell' what benefits might emerge from these experiments, it does not follow that we have a justification for engaging in them, and talk of the need for 'basic research' will not supply the deficiency. The fact is that if we take seriously the moral demands of animal welfare (as in my view we ought), we will have to meet the requirement that there is a clear causal connection between experiments such as these and reasonably anticipated, demonstrably beneficial results. Such a requirement, I am inclined to think, is relatively rarely satisfied at present. Most of these experiments are just that – purely experimental – and the suffering of animals that ensues from them, consequently, awaits any proper justification. Of course we can say (as the scientists involved generally say) that in the end it will not have been in vain. But as things stand we do not know this. Unhappily, despite many public protestations to the contrary, the suffering of animals for putative human benefit is still undertaken too lightly.

The 'slippery slope' and the 'sanctity of life'

The position is different when we turn to experiments on human genetic material. The focus of discussion here is embryo research, and whatever else there is to be said about it, a case cannot be made against it on grounds of unnecessary suffering, for the obvious reason that there is no suffering involved. Human embryos at the earliest stages of their development do not feel. Where then might an objection lie? This is one context in which people appeal to 'the slippery slope argument'. What is this 'argument' exactly?

The idea at work in the 'slippery slope argument' is that if we permit or engage in one kind of action, our doing so makes it more likely that some further action, to which we might otherwise have objected, will be undertaken. Now we need to distinguish here between logical and psychological 'slopes', so to speak. If the slippery slope against which we are being warned in the present context is a *psychological* one, it needs to be substantiated with empirical evidence in support of the claim that human beings who use embryonic material for research purposes are more likely to do other plainly objectionable things to human beings. This is evidence we do not have. Of course, similar sorts of claim are frequently made – that human beings having killed once will be more likely to kill again; that those who are willing to injure other people are more likely to murder them; that homosexuals are more likely to be involved in the sexual abuse of children – and this makes it tempting to think that there is some basis for the psychological version of the slippery slope argument. But despite a common assumption to the contrary, none of these putative connections is either obviously true or obviously false; each of them is a matter of human psychology that can only be determined by empirical investigation, investigation that to date has (at best) been very imperfectly conducted and produced very uncertain results.

An alternative version of the slippery slope argument gives it a logical rather than a psychological interpretation. According to this version, if we permit one sort of action (X) we will have no (or less) reason not to permit a very similar sort of action (X_1), which, as things stand, we would not permit. Translated into the present context what this means is that, even if we have no reason to suppose that those who are permitted to experiment on embryonic human material are likely to move on to experimenting first on foetuses and then on babies (the psychological version of the slippery slope), we will nevertheless find it harder to make the case against experiments on human foetuses if we have not objected to research on human embryos.

But once again I cannot see that the slippery slope depicted here is a real one. If there are solid grounds against permitting experiments on babies (as we may assume there are), these remain objections irrespective of the rights and wrongs of research on embryos/foetuses. If the objections to the use of human foetuses for research apply *pari passu* to the use of human embryos, then we have reason not to permit embryo as well as foetal research; if they do not, then the fact that foetuses cannot properly be used for research purposes does not automatically imply that we ought to forbid the use of embryos. In short, the objectionableness of X_1 cannot automatically rule out X. If this is correct, it follows that we are still in search of an objection to the research itself and cannot make effective appeal to slippery slopes down which it will lead us.

A familiar alternative to the slippery slope argument against genetic research on human embryonic material is that it violates the 'sanctity of human life' principle. As before, this principle is rather hard to state without begging the question. To put the point plainly, consider this question: does the 'sanctity of human life' rule out all killing in war? If we say that it does, we have not in fact given a reason against killing in war, but rather declared it wrong by *fiat*. If, on the other hand, we say that the sanctity of life principle does not automatically entail pacificism (*some* killing in war may be permissible), then we leave the rightness or wrongness of wartime killing undecided in the light of the principle. For example, killing aggressive enemies in war as an act of collective self-defence is clearly taking human life intentionally. We may, if we wish, declare it to be a violation of the sanctity principle. Even so, since it is obviously not in quite the same category as the aggressive killing of the innocent in war, *either* our declaration simply generates a reason not to accept the principle, *or* it acknowledges a (non-pacifist) application of the principle compatible with drawing a distinction between the two cases. But in this case we have rendered the principle itself silent on the question of where exactly the relevant moral difference lies.

This same difficulty arises for those who appeal to the sanctity of human life in order to rule out research on human embryonic material. Either they simply *assert* that the principle applies in this context (in which case those persuaded of the benefits of such research can equally simply reject the principle) or else they acknowledge important moral differences between embryo research and (say) the sort of research conducted by the Nazis, in which case appeal to the principle cannot settle the issue. In fact those who invoke the sanctity principle are in a weaker position here than in the wartime killing case, because they have to show that the sanctity of human life can be violated not only by destroying it, but by treating it in certain ways, i.e. using it for the purposes of experimentation. This important extension of the principle is not easy to sustain. While there is clearly *some* element common to all killings in war, whether offensive or defensive, there is *no* evident common element between these and the use of embryonic human material in the laboratory. Killing other adult human beings is always a serious business, even when carried out in self-defence. What we have yet to see is why we should regard embryonic research in this light, and merely invoking something called 'the sanctity of human life' cannot provide us with a reason.

Supporters of the sanctity principle are likely to think that it has been dealt with too speedily here. And there is indeed more to be said. The 'sanctity of human life' is often taken (implicitly) to mean much the same as a rather different principle – 'respect for persons'. Now these are quite different principles and they are often confused. But this general confusion does not matter very much so long as we are clear about just which idea is operating in any particular context. 'Respect for persons', stated at its most abstract, is the view that there are things that may never be done to people, irrespective of the desires or benefits to others, including the general welfare and the democratic will. The standard example is slavery, of course, but torture is also a plausible forbidden. Slavery can never be sanctioned, even though there are contexts in which it could be shown to be for the general good and/or in accordance with the will of the majority, democratically expressed.

Respect for persons is a principle that the modern world has ceased to contest directly, though arguments still rage about its precise scope and implications. However, in the case that concerns us here – the use of human embryos for the purposes of genetic research – no such arguments need rage, because there is very good reason to hold that human embryos are not persons in any sense whatsoever. Despite the fact that it conflicts with the teachings of the Catholic Church to which millions subscribe, this claim can be asserted with some confidence, it seems to me: but only if we sharply distinguish it from the very much more vexed issue of whether *foetuses* are persons. While there is a plausible case to be made for the claim that even at a relatively early stage of development a human foetus is identical with the adult that it will become, this is *not* true of the early embryo that precedes it. For the first fourteen days the embryo is most properly described, not as a distinguishable individual, but as a collection of cells. Differentiation into discernible parts of the body has not started, and crucially, the embryo can split into twins – or triplets or quads. This fact generates a compelling argument, which runs as follows.

Genetically identical twin adults, Jack and John say, are different persons, though they have the same originating embryo x. Now if we say that the embryo is itself a person, it can only be the person that develops into the resulting adult. But then we must say *both* that Jack is identical with x, *and* that John is identical with x. The concept of identity requires us to conclude, in this case, that Jack is the same person as John. But Jack is *not* the same person as John; they are distinct adults. It follows, therefore, that the attribution of personhood to x generates a contradiction. Either this, or we say that x is a *third* person, one who goes out of existence when the foetus appears. This in itself is very odd, of course, but since 60–70 per cent of embryos never implant and something in the region of 20–25 per cent miscarry, the claim that for these fourteen days an individual person exists implies that the vast majority of persons never make it past this stage. Could it be true that the countless millions of people who have lived are only a small proportion of all the people that there have been, the majority not having survived past fourteen days as an embryo? There is no contradiction in asserting this, but it seems a high price to pay for avoiding one.

To summarize once more: neither the slippery slope argument nor the sanctity of life principle, even when interpreted along the lines of 'respect for persons', can supply a clear and compelling reason for vetoing research on human embryos. But if the matter can be so easily settled, why is there so much concern? The answer, I think, is that other important considerations surround the question of research on human embryos. Moral doubts arise not only about the thing in itself, but about the culture it might give rise to. This too is a kind of slippery slope argument, but of a rather different sort. It is not so much a downslide that is anticipated as a spill over into other areas of conduct. Chief among these concerns, I think, is the question of how such material would and should be obtained. Should scientists be restricted to using spare material that arises from the course of other investigations? Or can they deliberately create and 'harvest' human embryos, so to speak? Is it permissible for human beings to buy and sell embryonic material? Could a woman rightly take employment (and be employed) as a supplier of embryos? It is questions such as these that most usually create moral alarm about human embryo research, and persuade those who contemplate it that the whole subject is a moral hornet's nest into which we should not allow ourselves to stumble.

Now if this is indeed the major source of our unease, it may be worth noting that it is not without precedent, and that the precedent in question is instructive. In the nineteenth century, Burke and Hare, working as body snatchers, provided the laboratory anatomists of their day with human research material (corpses in this case) in a way that opened them up not only to moral opprobrium, but to legal prosecution. It is to be observed, however, that at that time the anatomical investigation of corpses had become legal and was generally regarded as a morally acceptable practice; it was the culture of the body snatchers that was to be deplored. This fact was itself the outcome of an important change. At earlier periods, anatomical investigations of corpses had been both illegal and expressly condemned by the Church. (The anatomy theatre in Padua had a false floor so that both anatomist and corpse could be hidden from the authorities at short notice.) This negative attitude slowly changed to one so positive that it became (and remains) morally praiseworthy to assist the cause of medicine by leaving one's body for research. But if we praise those who give their bodies for research, why condemn those who give embryos? Nor is the case restricted to the self-giving of whole (dead) bodies. Equally praiseworthy nowadays is the giving of blood, bone marrow and kidneys by the living. Why should one part – embryonic material – be so special? It is hard to see that it could, unless we hold the embryo to be a distinct individual, which is the contention that the first argument refutes. By contrast, it is easy to see that the contemporary scruples about the use of human embryos are strikingly like the sort of moral squeamishness that was felt in previous centuries about the dissection of human corpses, a squeamishness that was finally overcome to the great and lasting benefit of humankind. Yet even when these clear and telling parallels are rehearsed, contemporary anxiety does not much abate.

This is a fact that brings us to a further range of topics. If what I have been arguing in this chapter is correct, then genetic engineering, far from being novel, is centuries old, and even cloning is not new. The strongest environmental warnings against genetic 'tampering' derive their strength from a 'precautionary principle' that it is either irrational or practically paralysing to invoke. As far as the thing in itself is concerned there is no obvious slippery slope (either psychological or logical) on which it launches us, no conception of 'the sanctity of human life' that it demonstrably violates. And now we have found that the 'spill over' effect that people fear is not so very different from earlier fears which we in our time have seen to be ill-founded and, where allowed to prevail, retrogressive. If all this is true, what can be left to sustain caution and anxiety except ignorance, prejudice, superstition, and irrational fears?

This is a question of the first importance because only if we answer it can we make sense of the attitudes of very large numbers of people. In my view we will not find another, hither undisclosed and decisive objection to genetic engineering; this chapter has aimed to examine the most plausible and has found none of them compelling. Rather if we want to understand the basis of the ambivalent contemporary attitude to genetic engineering we will have to return to the context of Chapter 1 – the place of science in society – and to consider further some of the themes of Chapter 2 – the conflict between Darwinism and religion. The starting point, however, is to look a little more closely at one of the ideas that we have had occasion to invoke in this chapter – the sanctity of human life. What exactly does the word 'sanctity' mean here? What could it mean?

4
PLAYING GOD

Secular versions of the sacred

It is a striking fact that moral debates conducted in the secular spirit of the (largely) post-religious world of contemporary Western society often continue to employ concepts whose original and natural home is a religious one. The 'sanctity of life' is just such a concept. 'Sanctity' means 'holiness' or 'sacredness'. What application could it have, then, in a world of ideas that has abandoned the sacred? I say 'a world of ideas' because it is evident that in the real world, which is to say the world of beliefs people actually hold and practices they still engage in, religion is alive and well. Recent estimates, for example, suggest that Christianity currently has somewhere around 1.9 billion adherents and Islam somewhere around 1.2 billion. The former figure represents approximately the same proportion of the world's population as in 1900 (one-third), and the latter a considerably larger proportion (perhaps three times as many). Nor are all these people to be found in non-Western (or non-Westernized) countries; as I noted in the opening chapter, the United States stands out among the industrialized nations as one in which unrivalled economic prosperity and technical advancement are combined with a high level of religiosity. Arguably it is only in Western Europe in the second half of the twentieth century that popular subscription to and participation in traditional religion has declined markedly (though even here a 'new-age' spirituality has its adherents). Accordingly there is a real danger that our understanding of religion in the modern world should be distortedly Eurocentric. Perhaps the European experience, rather than being a harbinger of the way the world in general will go, is a misleading aberration.

But whatever may be true about the cultural status of religion in the real world, to most minds its diminishing importance in the world of ideas, where Europe has been particularly influential, is evident. In the debates between Darwinism and creationism, discussed in a previous chapter, the general assumption on the part of informed participants is that creationism is the resort of cranks and dogmatists. Although, as I pointed out, there is considerable ground for thinking that dogmatism is a notable feature among some neo-Darwinians as well, this does not much alter the general perception that they have something called 'science' on their side, while the religionists have nothing better than a personal 'faith', or a rationally ungrounded belief in biblical or koranic revelation, or in papal authority.

This is not the proper place for a full discussion of these matters (which in any case I have addressed at length elsewhere: see Graham 2001). The point to be made here is that despite this cultural superiority, there are contexts and occasions when the authority of scientists is brought into question. Once the debate turns from genetic explanation, where 'science' rules supreme, to genetic engineering, where 'ethical' issues are thought to arise, the image of Frankenstein begins to overshadow that of Einstein, and even the secular world of ideas tends to reach for concepts that by its own assumptions are outmoded. It is doubtful, in my view, whether there is much grounding for a belief in the 'sanctity' of life in traditional Christian doctrine. In so far as there is, it draws upon the idea that human beings are made in the image of God, and this, evidently, requires there to be a God whose image it is. In the mind of the secularist, by contrast, all appeals to God and the supernatural have been abandoned. How then can any conception of the *sanctity* of human life remain?

A similar point can be made about another conception, namely 'playing God'. This is an idea regularly invoked by those who otherwise have no use for theological language. (*Playing God* is the title of an early book on the genetic revolution by June Goodfield in which there is no reference whatever to theology.) Yet if, as the secular world believes, there *is* no God, how could there be any danger of human beings illegitimately abrogating to themselves His function? In fact the position is worse than this. For many secularists it is not just that something that might have existed does not; rather they think that the very idea of God is meaningless. But in that case any anxiety about *playing* God must be meaningless also.

Jonathan Glover has made this point.

> The objection to playing God has a much wider appeal than to those who literally believe in a divine plan. But, outside such a context, it is unclear what the objection comes to. If we have a Darwinian view, according to which features of our nature have been selected for their contribution to gene survival, it is not blasphemous, or obviously disastrous, to start to control the process in the light of our own values.
>
> (Glover 1984: 46)

Glover himself does not give much credence to the idea of playing God, a view others share. John Harris, for example, dismisses the idea in a single paragraph, declaring it to be 'a non-starter' (see Harris 1998: 178). And even those whose orientation is theological tend to be negative. In the conclusion to his book *Playing God?* Ted Peters writes: 'After trying to discern whatever theological content the phrase playing God might have, it appears that it has very little', though he thinks that 'as a warning against a foolish prometeanism we should heed it' (Peters 1997: 177).

Yet if it is plausible to hold that, strictly speaking, there cannot be any secularized version of the sanctity of human life or the *hubris* of playing God, such expressions notably fail to go away. What are we to make of this? My own view is that the use of these residual phrases aims to capture something which, it is believed, is of fundamental importance in understanding and ordering the relations of human beings to the world around them. The phrases themselves may be largely otiose, but they point to ideas that, even in a secular world, it is hugely important not to ignore or neglect. So runs a common line of thought. The question, of course, is what these ideas, shorn of their theological overtones, might be, and whether it is indeed important not to overlook them.

Consider first the sanctity of life. To uncover any valuable version of this something more needs to be said about its religious origins. The distinction between the sacred and the profane runs through all religions. In contrast to the profane, which is the character of the larger part of human life, the sacred is that which cannot be approached without 'fear and trembling', a dimension of experience famously described by Rudolf Otto as '*mysterium tremens*', the proper response to which is awe and humility. So conceived, the sacred is an area 'where angels fear to tread'. If so, since angels are superior to us in every respect, it must be even greater folly for human beings to tread there. But in the absence of God and His angels, what else could erect such a rationally insurmountable barrier?

Genetic trespassing

One answer is 'genetic trespassing'. While the expression itself is attributable to Mae-Wan Ho, the idea that there is something illicit in crossing species boundaries has recurred regularly in the literature surrounding genetic engineering. Richard Sherlock has usefully distinguished three versions of it (Sherlock 2001). The first is that the boundaries between species are naturally established in some way, and that to take some of the genes of one species and place them in the genome of another is to cross this boundary and hence against nature. Clearly, there is at work here a normative concept of 'the natural'. As normative concepts, 'natural' and 'unnatural' are notoriously hard to explicate satisfactorily. For that reason, perhaps, the negative term has fallen into disuse. (Consider the 'old-fashioned' connotations of the phrase 'unnatural practices'.) But the positive term has always retained some currency, as in the expression 'natural rights' and 'natural parenting', and has recently gained sufficient prominence to make 'natural' a powerful word of commendation, one extensively exploited in the food and clothing industries, for example. Yet the difficulties in sustaining 'natural/unnatural' as a normative criterion remain. In so far as it means 'what is found in nature', it seems that genetic trespassing goes on all the time, because the boundaries between species are not fixed and final. Evolutionary history shows them to have been rather fluid, in fact, a fluidity further revealed in the familiar practice of cross breeding animals, which human beings have engaged in for centuries. And cross breeding is just a 'low technology' version of genetic engineering. More importantly, cross breeding can be found taking place spontaneously and without human intervention. In the case of animals this often, though not always, leads to sterility, but the same is not true of plants (which is why the definition of distinct species is rather easier in the case of animals than of plants). If interpreted in this way, 'genetic trespassing' *is* a fault, but it is a fault deeply embedded in the natural world and not a phenomenon that has newly arisen with the arrival of biotechnology.

A different interpretation restricts genetic trespassing to those cases in which the act of 'trespass' carries large-scale dangers. Once more, the relevant issues here have largely been discussed already, in the examination of the 'precautionary principle'. Let us agree that some crossings of species-specific divisions are relatively innocent, while others are much riskier. Since risk comes in degrees, we cannot separate out two exclusive classes, the one permissible, the other impermissible. And since, to repeat an earlier argument, a wide variety of human actions (perhaps *all* actions) have degrees of risk attaching to them, we have no choice but to investigate the matter case by case and make, so far as may be possible, the appropriate estimate of costs and benefits. The point about the precautionary principle is that it tries to circumscribe some risky actions in a way that rules out cost–benefit analysis *a priori*. But for reasons that need not be repeated, there is no rational justification to this restriction. Difficult though it may be in practice, as unquestionably it often is, there is no theoretical boundary to the cost–benefit analyses that we can, and ought to, engage in.

Possibly the most interesting interpretation of the idea of genetic trespassing is that which emphasizes the speed of genetic change. The proponent of this interpretation will agree that natural evolution includes many instances of specific boundaries being crossed. However, this happens very slowly, and its slowness allows for the gradual adaptation of the emergent 'new' species to its environment. The result is that the impact upon that environment does not run the risk of being cataclysmic. By contrast, the sort of boundary crossing that modern biotechnology makes possible is immediate. As a result, new types of creature are put into an environment that has had no time to adapt to them, with the likely, if not inevitable, result that the delicate ecological balance of that environment is drastically affected.

On first appearance this seems a plausible line of thought, and thus a telling objection to contemporary ambitions and practices. Yet it is difficult to know exactly what sort of claim it is. Is it an empirical claim based upon evidence of previous experience? Or is it a speculation based upon the concept of 'speed' that is precisely formulated to capture a certain kind of case? If 'speed' is defined in terms of *immediate* genetic change (as opposed to the same change brought about by a slow and gradual evolution) it seems to be a strictly empirical question whether it proves specially harmful or not. Since it seems perfectly possible that there should be immediate genetic changes that are very small – at the level of bacteria for experimental purposes for instance – we cannot predict in advance and *in general* that all such changes will be dramatically harmful. If, on the other hand, we define genetic 'speed' in terms of the scale of its impact, we have reverted to the second interpretation, and the arguments brought to bear against it apply once more. We should avoid every sort of genetic 'meddling' which carries a great risk of widespread environmental damage. No one seriously disputes this, I imagine, partly because the concept of 'damage' is normatively negative. It is a sensible principle with which to operate. What we need to observe, however, is that it is not one that automatically captures (and hence rules out) 'speedy' genetic modifications. Nor is it a principle to be invoked exclusively in this context. Major environmental damage is just as much to be avoided when it arises from industrial pollution, or urban development, or agro-business, as it is when it arises from biotechnological innovation.

The concept of 'genetic trespassing', I am inclined to think, is yet another failed attempt to produce a secular version of a religious idea – that the order of the world has been established by the will of God, and that the place of human beings is to acknowledge and respect the limits and boundaries set by the divine will. As a matter of fact, if we pay serious attention to the evidence of natural evolution, even working within this religious conception it seems that God has not made the divisions between species one of those boundaries, and has (very occasionally) permitted reproduction by cloning (witness the strange case of the whiptail lizard of Colorado which breeds by the fusion of two female ova).

In any case, the concept of genetic trespass, even if it could be made coherent and compelling, would not give us reason to focus on the context in which most people fear that genetic engineers have taken to 'playing God'. This is the context of humanity itself as a species, and in particular the ambition of fashioning our own nature. Whatever the rights and wrongs of creating new types of bird, or plant, or fish, about which attitudes can be more and less relaxed, the idea of setting about the alteration of humanity itself raises much more pressing anxieties. If genetic trespassing is an objection that cannot be made to stick, is there another that might be more successfully restricted to this limited context?

Rights and equality

One supposedly morally insurmountable obstacle to many of the things that human beings attempt is 'human rights'. The concept of human rights can be so construed as to supply us with a secular version of the sanctity of (human) life. Human rights constitute a moral boundary that it is always and everywhere wrong to cross. What is the basis of this boundary? One explanation is equality. Human beings are fundamentally equal, and on this interpretation the principle of the 'sanctity of human life' is simply another way of affirming the fact. But in what respect are human beings fundamentally equal? Not in endowment or accomplishment, certainly. No one can argue with any plausibility that human beings are equal in their talents or in what they achieve. Shakespeare had a gift, and was productive, to a degree that will almost certainly forever outshine the efforts and abilities of every other poet and playwright. Alexander the Great made a mark upon his times, and upon history, far exceeding that of any of his contemporaries, and the majority of political leaders since. Thomas Aquinas was an intellectual luminary with very few counterparts then or now. J. S. Bach was a composer surpassing nearly every other. Bobby Jones was a golfer of quite exceptional grace and skill. It seems that as far as talent and achievement are concerned, human beings are most decidedly *not* equal.

Such being the case, the assertion of a fundamental *equality* between human beings must precisely seek to deny that their evident inequality of ability and accomplishment is the ultimate mark of their significance. What is it, then, that the most and the least talented and accomplished have in common? The religious can provide an answer, of course. All, by their account, are equally beloved of God and equally in need of His mercy, by which alone they can be saved from perdition. If this is not a line of thought we can pursue, as the secular world holds, what is the alternative? I shall explore what seems to me the most promising – that no one is in a position to decide that the life of another is not worth living.

This is a powerful idea, even if in the end it falls short of the religious conceptions it aims to replace. It is connected with the idea of rights in the following way. Everyone has a right to life because no one is in a position to declare the life of another worth nothing. In the final analysis, the thought is, the value of a life is to be determined by the person whose life it is. Those who are able-bodied, intelligent and accomplished may, correctly, suppose that their lives are better than the lives of those who are handicapped, feeble-minded, sickly and inept. I say 'correctly' because there can be little doubt that handicap, feeble-mindedness, illness, and so on, are disabilities to be wished away. But they are not such as to allow us to declare the lives of those so confined to be ultimately worthless. 'It were better I had not been born' is a judgement to be made (if it is to be made at all) in the first person only. A large part of the evil of Nazism (and other such codes) lay in the fact that it believed to the contrary, and held that the (imaginary) 'Aryans' are in a position to pass ultimate judgement on the worth of other lives – Jews, Slavs, gypsies, the mentally retarded, and so on.

We might usefully draw a familiar distinction here. The lives of the healthy, the intelligent, the handsome and the talented are better lives than those of the sickly, the intellectually weak, the deformed and the talentless, but they are not *morally* better. There is a profound error in any assumption of moral superiority on the part of the physically and intellectually superior because, as I shall put it, an essentially first-person judgement – 'This life is not worth living' – cannot be made in the third person. ('Lives not worth living' is an expression used by the Nazis; see Nelkin and Lindee 1995: 171.) Here, I think, we do find something of a basis for a secular version of the 'sanctity of life' and a basis also for secular anxieties about 'playing God'. Those who do not respect the sanctity of life are assuming that their third-person view of the matter is authoritative, whereas only the first-person view can decide this. Whether my life is worth living or not is a matter for *me*, exclusively. Similarly, those who choose to play God (by genetically engineering certain outcomes over others) are assuming that they are in a position to decide, in this case in advance, that some lives are more worth living than others. And no one is actually in this position.

Whether, ultimately, we can make this secular account of the sanctity of life fully intelligible is a difficult philosophical question to which I will return. But for the moment I propose to work within this moral framework and see what can be said about two important topics that the advent of modern genetics (allied with other techniques) has given rise to – human cloning and designer babies.

Human cloning

As we saw in a previous chapter, cloning – the direct reproduction of a copy of an adult entity – is not new. Cloning of plants by means of cuttings has taken place for a very long time. What is new is the cloning of animals, and though it began with sheep and has largely remained confined to domestic animals, with the advent of such techniques it seems that the cloning of human beings cannot be far behind. What, if anything, would be wrong with this? Why is it not to be regarded as just another way of dealing with childlessness? That something would be radically wrong about human cloning seems evident to many people. Both governments and scientific bodies have been quick to urge a world-wide ban, and by and large, it seems, public opinion supports them in this. Now, if there is no denying that the instinct against it is strong, and highly influential, it is rather less clear that there is something against it other than instinct.

One preliminary observation worth making is that responsibility for launching us upon a morally unnerving path does not lie with genetics alone. The technique of *in vitro* fertilization, whose initial purpose was to overcome certain physiological barriers to child bearing, is hugely important in bringing about the possibility of human cloning. This is further evidence, in my view, that science and technology are in important ways independent of each other. Without the relatively autonomous technology of *in vitro* fertilization, the science of genetics would not have presented us with the problem of human cloning. And it is the extension of technology more than the fresh discoveries of science that has brought the prospect of human cloning firmly into view. At any rate, the cloning of human beings is now not just a theoretical, but a technical possibility. And given the well-publicized intentions and efforts of some scientists and medics, it is a possibility that will soon become a reality. Indeed, as I noted earlier, some scientists have asserted with great confidence that the first human clone will have been born within a short time of this book's being published.

What is there to be feared about this prospect? An obvious response is that the confidence is misplaced, or at any rate misconceived. While it may be true that all the scientific and technical pieces will soon be in place to make an attempt at human cloning, it is also true that there is every chance of such an attempt's having highly undesirable consequences. The cloning of humans is an extension of techniques developed for the cloning of animals, and whereas the cloning of plants is simple, the cloning of animals is not. Since Dolly the sheep made her first appearance in 1996, it has become evident (as I noted in the case of ANDi, the genetically modified monkey) that present techniques for animal cloning are very imprecise and wasteful; hundreds of attempts can be needed to secure a single successful outcome, and the production of deformed and defective creatures along the way is common. Now while this may be acceptable with respect to sheep (it is not an issue I shall discuss here), it is hard to see that it could be acceptable with respect to human beings. As things stand, anyone who attempts to clone a human being will have to create multiple foetuses in the knowledge that most of them will not survive, and that of those that do there is a high chance of some having brain and other defects. In fact, recent research suggests that the technique of cloning currently deployed may be counterproductive, in this sense: the method itself may induce abnormal gene expression, with the result that the animal produced is not a replica of the adult, but a modified version of the original. In fact we cannot rule out the possibility, even, that the method results in genetic damage. Moreover, such is the inefficiency of the method that the modification can result in defects and deformities so bad that the creature has to be destroyed immediately.

Clearly, such a process applied to human beings would be intolerable. No one could suppose that the hit-and-miss production of human beings, and the destruction of those who came out of the process deformed, is morally acceptable. What ends are there that could possibly justify this? Childlessness is not such a huge burden that its relief could outweigh such horrors, and potential future medical therapy benefits are both too remote and too speculative to provide any serious justification. In any case the scenario seems far too gross to admit of such cost–benefit calculations – the contented parents of a healthy child on one side, and a collection of human deformities on the other. For the foreseeable future, therefore, it seems that morally speaking the cloning of developed human beings is out. (The cloning of human embryos for the purposes of research is not ruled out by these considerations and is a topic to be returned to.)

This is a significant conclusion, in my view. Sometimes people speak as though the practical and the moral were to be thought about in different ways. Yet morality is to a large extent about action, and practical considerations may make actions immoral. When this is the case, impracticability – the impossibility of securing a morally permissible or morally desirable outcome – settles the matter. If the former, we are wrong to attempt it; if the latter, we have no obligation to attempt it. ('Ought implies can' is an ethical principle long acknowledged.) But impracticability does not put an end to every argument, and many people, even when they are convinced that we are not faced with the immediate prospect of human cloning, designer babies and the like, still want to know whether there are some 'in principle' objections to technological experiments that are currently impracticable. Moreover, their desire to know this is often a function of their awareness that practicalities can change. This is especially true in the context with which we are concerned here – biotechnology. Many of the things that are now common practice were not merely unknown but inconceivable a relatively short time ago. The inefficiencies of present animal cloning techniques are thoroughly familiar to those who deploy them, but a major scientific effort is already underway to understand precisely why things go wrong, as a preliminary to improving their efficiency. Consequently, to rest content with the argument that, as things stand, any attempt at human cloning is likely to go wrong, says nothing about the theoretical case in which it goes right. Moreover, though at present the contingent fact of its gross inefficiency rules it out, morally speaking, it does not rule out efforts to improve that efficiency, efforts that are as a matter of fact taking place. Suppose that these succeed, that the technique of animal cloning becomes far more precise and effective and its extension to human beings very much less likely to go wrong. What then? We may be reasonably confident, as I am, that this day is very far off, and may never come. Nevertheless, the inefficiency argument rests on contingent facts, and it is in the nature of contingent facts that they can change. Is there anything to be said against human cloning *in principle* that would make it morally repellent however good the technique might become?

A great many people believe so, on a variety of grounds, but there is at least one objection to human cloning that is easily dispelled. In public debates and newspaper articles on the subject, the idea of 'human clone' and 'zombie' are often confused. Zombies are to be defined as slave-like creatures whose behaviour is determined and controlled by those who create them. There may never have been any zombies, but if there were, evidently, they were not clones. Conversely, if the time comes when there are human clones, there is no reason to suppose that they will be zombies. In so far as a human clone is to be characterized as a person genetically identical with some other person, we do not need to turn to the marvels of modern biology to find examples. Identical twins fit this description. But it goes largely without saying that identical twins are no one's zombies, least of all each other's. Indeed, the phenomenon of identical twins is one worth dwelling on in this context. Though genetically identical, such people are not indistinguishable from each other, even in appearance. And when it comes to personhood, the identical twin has as much individuality as anyone else. If identical twins are natural human 'clones' without any element of the zombie, there is no reason to suppose that future technological clones will be zombies either.

The fear that clones will form the basis of an unthinking slave-type class is without foundation, then. What other fear might there be better grounds to entertain? One seemingly significant difference between 'natural' clones (i.e. identical twins) and artificially contrived clones lies in their respective ages. While identical twins are of the same age, cloned children would be genetically identical with adults several years their senior. Might this constitute a problem? Some people think it does because it seems to make the clone a mere replication of the adult, and hence not an individual in his or her own right.

On further reflection, though, it is hard to see that age difference in itself could raise any special difficulty. If the anticipated loss of individuality is a version of the 'zombie' problem, there simply is no reason to think that I (the clone) would be any more the slave of the originating adult than I would be the slave of my identical twin. Of course it *could* be true that the adult from whom I am cloned does indeed dominate and enslave me, and it may be that the ability to do so is partly a matter of being older. But this possibility fails to mark out the cloner–cloned relationship in any special way. The same unfortunate features can be found to blight and distort relations with natural parents also. Unhappily, those who use their greater age to bully and domineer are not a phenomenon about to arrive with the advent of human cloning.

Alternatively, the fear might be interpreted as arising from the belief that replication destroys personal individuality *numerically*, so to speak, by making the person no longer (literally) unique. The first point to be made about this idea, if there is indeed anything to it at all, is that it works both ways; the adult would be as much a loser in the individuality stakes as the clone. The second is that identical twins do not cease to be individuals because there are two of them. If we do not think this a problem in the case of identical twins, why would clones present us with it? Indeed, in this respect artificial clones are, if anything, on firmer ground than identical twins because, given the difference in age, their life story is guaranteed to differ from that of the adult from whom they were cloned. Identical twins could (theoretically) have the same upbringing, education and so on, and hence an identical pattern of personality forming life experiences, and as a result (perhaps) almost indistinguishable personalities. This simply will not be true of those who are cloned from existing adults.

Any anxiety about creating either zombies or 'mere' replicas seems ill founded. Perhaps a better way of getting at the moral issues involved in human cloning would be to look at it from the point of view of the cloner and ask why anyone might want a clone of themselves. One answer, to which I have already referred, is childlessness. It is easy to imagine circumstances in which a couple cannot have children by natural means, and for whom the more familiar applications of *in vitro* fertilization prove unsuccessful. Many of those who advocate human cloning argue that were cloning to prove possible in these cases there would be just the same reason to approve it. Is this true?

We need to leave aside here the 'slippery slope' arguments that are frequently invoked. I have tried to answer these in the previous chapter, but it is worth repeating that in their most familiar expositions they generally appeal to ignorance – who knows where all this will lead? Though often rhetorically compelling, such appeals to ignorance are intellectually weak because the ignorance they appeal to cuts almost every way: who knows what all this will *not* lead to? If the dangers are unknown, so too are the benefits. Some other argumentative strategy is required to settle the matter.

There are, as it seems to me, three distinguishable reasons for seeking human cloning. Perhaps the most commonly advanced justifies it as a way of producing embryos whose stem cells can be used in medical therapies of one sort or another. The second is the reason just alluded to – to alleviate the distress and suffering of childlessness. The third is the rather more ambitious – to seek to reproduce a specific type of being – designer babies in short.

Stem cell production

One of the most recent, and most impressive, biomedical technologies is the use of stem cells. Stem cells have a special property – they are undifferentiated. That is to say, stem cells have not taken on the special properties and functions of liver cells, heart cells, skin cells, and so on. But they can become differentiated, and take on these properties. This makes them dramatically useful. Stem cells can thus be used to repair organic damage, to recreate parts of the human body that are diseased or malfunctioning. Thus they present us with wonderful new therapeutic possibilities, several of which have already been impressively demonstrated – bone marrow transplants to regenerate a healthy blood system in patients with leukaemia, for instance.

But to realize these possibilities we need a supply of stem cells.

Where are they to come from? One current source is aborted foetuses. Another is adults, in whose marrow, especially, stem cells are to be found. Another is human embryos, and though we can use embryos for this purpose without cloning them, it is the use of cloned embryos that has prompted the most debate. Why? Actually, it is a debate that some people think we can avoid. We can avoid it, they allege, because there is no need to produce stem cells from embryos at all, cloned or otherwise; stem cell production from volunteering adults is quite sufficient. Moreover, some of these same people allege, the technique of de-differentiation, which scientists are beginning to investigate, overcomes the limitation that has hitherto been thought to operate. That is to say, if the supply of stem cells from adults is limited, the supply of ordinary cells is not. And if these ordinary cells can be manipulated to lose their specific characteristics – to be de-differentiated – then we have as ample a supply of stem cells as we could want without entering upon human cloning even at the embryonic stage.

Everyone will agree that embryonic cloning is unnecessary for the production of stem cells for the purposes of medical therapies. The question is whether it is especially advantageous. This is clearly a technical issue. Seemingly equal expert opinion appears to be in considerable disagreement about it, and it is plain that no philosophical or ethical argument can settle the matter. But the significance of the debate turns on whether it would be better, if we could do so, to seek a source of stem cells from adults rather than from cloned embryos, and this in turn implies that the cloning of human embryos would be *prima facie* objectionable, that is to say, something which *could* be justified in terms of medical benefit but which certainly *needs* justification. Is this true? Should we think that the cloning of human embryos is to be avoided if it can be?

It is sometimes argued, and more often assumed, that the use of cloning for therapeutic purposes is tantamount to creating one person solely for the sake of another. This contention invokes (some version of) the principle of the fundamental equality of human beings. In the previous chapter, however, I argued that in the distinguishable stages of the development of a human being, there is an important gap between embryo and foetus, a gap, moreover, that is much more evident than the gap between foetus and neonate. Whereas it is possible, plausible even, to argue that foetus and neonate are equally stages in the life of the adult person, this cannot be argued for the embryo. As the existence of identical twins demonstrates, there is no reason to identify the embryonic bundle of cells with any given adult. If this is correct, it follows that there is in fact less reason to object to the use of cloned embryos for therapeutic purposes than there is to object to the use of aborted foetuses. Given that a foetus is one stage in the life of the adult it may become, there is at least an argument to be made for the view that using an aborted *foetus* for the benefit of an existing individual is indeed determining the value of one human being in terms of its capacity to benefit another. But producing cloned *embryos* that never proceed beyond the embryonic stage, and using them for the health of the adult from which they are cloned is *not* a case of making one person solely for use by another. From this point of view, at least, the position is more like using a bit of myself (a vein from my leg for the repair of my heart, say) than using the same bits from someone else.

The collapse of this particular objection does not necessarily give the green light to cloning human embryos. It turns out, for example, that stem cells produced in this way may be unstable, and may even be cancerous, a tendency that adult stem cells may not have. Though the context of de-differentiation makes this difference highly uncertain, cloning for stem cell production may have dangers peculiar to it sufficiently serious for us to look for other ways in which the same therapeutic ends might be better accomplished. The issue, however, is essentially a technical one. We will all agree that a better method is to be preferred over a worse one. This is a quite general principle that makes no special reference to cloning or stem cells. Accordingly, it seems to me, to justify embryonic human cloning in terms of stem cell production does not have any grandiose implications. Though as I myself have argued, it does not involve the improper use of human beings (who happen to be at an early stage of their development), the benefits it promises are contingent and may in the end be more effectively obtained by other methods.

Reproductive technologies

This brings us to childlessness, the second of the purported justifications, and that most commonly invoked by the doctors and scientists who are enthusiasts for human cloning. Their argument, briefly, is that we are already thoroughly familiar with highly technical devices whose purpose is to overcome natural obstacles to the generation of children and if (leaving the technical uncertainties aside) cloning is simply another technique for doing the same thing in a more refined, reliable and effective way, it cannot be regarded as any more objectionable than the equally hi-tech methods that preceded it. This is a common inference – from the acceptability of IVF to the acceptability of cloning – and it seems reasonable. Yet it ought to be examined more closely, because the position is not as clear as one might expect. To begin with, of course, the inference supposes that there is an adequate justification for using existing techniques. This can be questioned. It is not a matter I shall address directly, but some of the points to be made about reproduction by cloning could equally well be made against *in vitro* fertilization (test tube babies). A telling question is this: What exactly is the condition of 'childlessness' that either technique aims to remedy? Within the general expression 'having children' we can distinguish three different things – being a biological parent, bearing children, and rearing children. The distinction between the first two and the third has been familiar for centuries; a mother who adopts *has* (in the sense of 'rears') children, without being their biological parent, or having given birth to them. Modern technology has introduced a novelty in that now the first two can also be separated – surrogate mothers can give birth to children to whom they are not biologically related, and conversely the biological mother of a child may not be the person who gave birth to it.

What this means is that, thanks to modern technology, the three senses of 'having children' can be differentiated physically and not just conceptually. What then is childlessness? If it means not *rearing* children, then adoption is an answer to it, and given that, thanks to wars, disease and poverty there are innumerably many children in the world available for adoption, we do not need to go the technically difficult and expensive road of test-tube babies and cloning to remedy the problem of childlessness. A more urgent, cheaper and more effective remedy would lie in the abolition of artificial national boundaries to adoption.

If 'having children' is defined so as to incorporate child bearing, then it follows that a woman who is the biological mother of children borne by a surrogate, whom she then rears to maturity is 'childless'. This seems absurd. The failure to bear children cannot, therefore, be a crucial aspect of childlessness. This leaves biological parenthood. But to make biological parenthood crucial is also odd, though perhaps less so. Imagine a case in which the sterile wife of a sterile man has a fertilized egg from anonymous donors implanted, and which she brings to term and raises to adulthood. It seems certain that in this case ordinary language would declare her to be mother of the child, and would continue to do so even if its anonymous biological origins became known. Plainly there is an interpretable sense to the idea that it is not strictly 'her' child, but it is not likely to be a sense that cuts much ice legally, personally or morally. What this case, in combination with the previous one, implies is that none of the three senses of 'having children' is a necessary condition of a meaningful use of the expression. The three conditions – biological parenthood, giving birth to, and raising to maturity – are of course present in the standard, natural case. But what modern technologies have done by making it possible to separate them, is to raise a doubt as to whether any of them is crucial. And if none of them is, it turns out that the non-technical means of adoption is as clear a case of 'having children' as the hi-tech alternatives.

It follows that 'childlessness' is a condition that can be remedied, and for a very long time has been remedied, by non-technological methods. Suppose we say, or rather, suppose a childless couple to say that adoption is not enough. Where might the missing element lie? It is sometimes said that people have a 'right' to have children. On a negative interpretation this is (relatively) unobjectionable. The claim means no more than that people cannot legitimately be *prevented from* having children. (There may be defensible exceptions, in fact.) But often the 'right to have children' is given a positive interpretation – that those who desire to have children are entitled to have their desire satisfied. This second interpretation can reasonably be disputed, however, because it is not at all clear that anyone could have the corresponding duty to ensure that this 'right' is realized. Who is it that owes it to a childless couple to bring it about that they have children? Even if such doubts are laid aside, we are left with the question of what exactly such a right would be a right *to*. In the light of the foregoing argument, it seems that there are several ways to 'have' children. If the 'right' is to rear children, this can be satisfied without any recourse to hi-tech methods. If it is to give birth (which could only apply to the mother, of course) this directs our attention to a certain kind of experience. Can anyone seriously maintain that there is a 'right' to have a particular kind of experience? Is it, then, that there is a right to be a biological progenitor? Again this seems to strain our intuitions. For one thing, we can be progenitors in total ignorance of the fact and even (thanks to modern technology again) progenitors *post mortem*.

Perhaps we should abandon the talk of rights, and speak instead of having children as a 'basic human good'. The concept of basic human goods is an increasingly popular one in contemporary philosophical discussion, notably in the work of Amartya Sen and Martha Nussbaum. The concept refers to those things without which a human life may be said to be incomplete. To deprive a human being of a basic human good is to do the person a deep injury, and to supply them with it is to benefit them greatly. Among the candidates for designation as basic human goods are the material means of survival, language acquisition (and a measure of education more broadly conceived), a sense of self-worth, and the development of innate talents.

Is the procreation and/or rearing of children among such 'basic goods'? Many people clearly think so.

> A popular book about infertility refers to the 'biological handicap' of those unable to bear children. The author interviews infertile couples who feel abnormal and diseased because they cannot bear genetic children. A fertility handbook conflates personal fulfillment, religious impulse and genetic destiny to explain the urgency to reproduce. 'Call it a cosmic spark or spiritual fulfillment, biological need or human destiny, the desire for a family rises unbidden from our genetic souls.' A character undergoing fertility treatment in a women's magazine short story

> says she and her husband are 'like warriors going into battle' against a body that has 'betrayed' her. And in *Resolve*, a newsletter on infertility, a writer describes the feelings of inadequacy and the lack of self-esteem among women unable to conceive. Having a baby is essential if a woman is to be whole, for the experiences of pregnancy and motherhood are 'the core of women's being'.
>
> (Nelkin and Lindee 1995: 62–3)

As Nelkin and Lindee point out, this attitude to childlessness, and with it a sense of the insufficiency of adopting, is a modern phenomenon. Earlier ages were much less intense on the subject, and viewed in this perspective, the contemporary obsession does seem excessive. While it can be said with certainty that having children, in the normal everyday way, is among the deepest satisfactions that most human beings enjoy, it cannot be said that a life without procreation is thus to be regarded as irredeemably blighted. In fact, and on the contrary, it is possible, and intelligible, for human beings deliberately to choose forms of life that expressly *eschew* procreation. In many cultures celibacy is an ideal, and widespread. Christianity and Buddhism are specially noted for the importance they attach to a life that renounces sexual activity, and hence procreation, in pursuit of other, spiritual goals, and whether we subscribe to such a view or not, the evidence seems to be that such a life can be a deeply satisfying one, more fulfilling, indeed, than many a parental life. This is why many Christians and most Buddhists, who are not themselves celibate, acknowledge and admire the special status of nuns and monks.

The possibility of childlessness as an *ideal* as well as a misfortune alerts us to a larger consideration which must enter into any appeal to 'basic human goods'. It reveals the possibility, and the need often, of making trade-offs between the various goods taken to be basic. The development of my innate talents, depending upon what they are, may require me to sacrifice a large measure of material advantage. Alongside the nun or the monk who trades sexual satisfaction for spiritual experience, is the artist in a garret, eking out a meagre existence for the sake of his or her art, a familiar, and far from despicable, or pitiable, figure. Both cases suggest that the 'rounded life', the life in which there is a reasonable measure of all the things that human beings tend to benefit from, may not be compatible with anything that might be called 'greatness'. The pursuit of the highest spiritual or artistic goals may require the renunciation of some of the things that in ordinary life are taken to be 'basic' goods.

In short, though the idea that having children is a 'basic good' is much more plausible than the suggestion that having children is a right (in the positive sense), it is a good that there can be reason to forgo, and the life that forgoes it is not thereby rendered poor and worthless. It follows that childlessness, however regrettable and, in the case of the couple who long for a child, lamentable, is not an instance of basic deprivation, like starvation or ignorance, which there is reason to take every possible measure to avert and avoid. Once more, there is a religious way of expressing this fact. Children are a gift, and it is in the nature of gifts that we neither have a right to them nor can we demand them. This is language that secularists not infrequently employ, of course. Yet, as with the sanctity of life and the risk of playing God, it is far from clear that they can do so meaningfully. How can there be gifts without a giver?

What should we conclude, then, about these hi-tech reproductive technologies? The answer seems to be that their deployment has to be placed in a larger context than that of the express desires of specific individuals or couples to procreate. We need to make a more general calculation that aims to determine the relative costs and benefits, social as well as economic, of the institution of *in vitro* fertilization or cloning as a standard social practice. In the previous chapter I stressed the importance of considering genetic screening in the same social context, that is to say, assessing its merits and demerits not just at the level of personal choice, but as something a society might go in for. We need to ask not simply whether, in specific circumstances, an individual might have reason to invest in the substantial cost of hi-tech remedies, but whether society at large, through the instrument of the state usually, has good reason to supply them. So, too, in the case of reproductive technologies. The question is not simply whether a couple who have the option of adopting should pursue the generally more difficult, and much more costly, option of overcoming natural barriers to the procreation and bearing of children, but whether there is some basis for thinking that 'society' should provide the means by which they may do so. The answer is 'yes' if, and only if, some fundamental right is violated or some basic good denied if it does not. And this, I have argued, is a case that cannot be made out satisfactorily.

There is a further point, which may be more telling with respect to human cloning than is the case with *in vitro* fertilization. Even if it is true that *in principle* there is nothing more to object to in cloning for purely reproductive purposes than in the technique of test tube babies, a technique that merely replaces normal processes of fertilization, it may nevertheless be the case that the larger impact of such techniques renders them objectionable. This is their impact on the pattern of human relationships. We do not know how human clones would as a matter of fact relate to those of whom they are the clones. It may be that cloned human beings would relate to their progenitors as natural children relate to parents. But this cannot be known. What can be known is that adoptive children and parents regularly encounter difficulties that natural relationships do not. Cloning is a radical departure from all hitherto known methods of procreation, far more radical than adoption, and much more radical than *in vitro* fertilization which for the most part results in the same outcome as procreative sexual intercourse. Prediction on this score is precarious, of course, though this works both ways, and I have already decried arguments from ignorance. Nevertheless, in the light of the radical nature of the departure, caution seems not unreasonable. If we add to these reservations the point about social provision, a plausible guess is that the amount of human distress which the availability of cloning is likely to eliminate will almost certainly be outweighed by the opportunity costs and inequalities that its introduction would bring about. If anyone thinks that cloning is a *simple* solution to an easily acknowledged problem, then they have reason to think again.

Such a conclusion, partly because it is to a considerable extent a matter of guesswork, is inevitably less than satisfactory. At best it introduces a measure of doubt about the merits of cloning as a reproductive technique, and further advances in the technology, together with an increase in general prosperity, could do much to alter the calculation. Is there nothing to be said in the abstract that might help us to arrive at a more definitive view on the issue? There is, I think, if we turn to the third kind of reason that people might have for resorting to the technique of human cloning – the desire not merely to reproduce but to replicate. This brings us to the topic of 'designer babies'.

Designer babies and eugenics

The cloning of plants, as I have noted several times, is a possibility with which we have been familiar for centuries, and the cloning of domestic animals, which is more recent, is not a matter solely of more efficient reproduction, but of the replication of preferred specimens. Though it may be true in the case of plants (not yet in the case of animals) that cloning is simply a quicker and more reliable way of reproducing healthy and productive varieties, this way of describing the matter leaves an important aspect out of account; these varieties are more *desirable*. Horticulturalists and agriculturalists *select* varieties, and the selection is made with a view to propagating the varieties that are more desirable from the point of view of human consumption, sometimes for food, sometimes for ornament, sometimes for pleasure. In so doing, they have no regard for the distinctiveness of the variety, still less the individual, that they select. A variety of grape, for instance, is selected in accordance with merits determined by the purposes of wine-making or fruit drying, and these are human purposes, of course. Similarly, the artificial fertilization of race horses or beef cattle (which the new techniques of animal cloning are, perhaps erroneously, expected to replace) takes its rationale from the interests human beings have in sport and in nutrition respectively. Now while quite unobjectionable in these contexts no doubt (though some would question this), there is an implication that is worth drawing out. All these activities suppose that human purposes are superior to the extent that they can rightly determine why and in what way other beings exist. The Oncomouse, discussed earlier, is a striking case, and implies further questions to be returned to. Applied to human cloning, though, it seems evident that a greater difficulty arises. Who do we suppose one set of human beings to be that they should determine the why and wherefore of the existence of others? Is this not precisely where the error of the Nazis lay?

To make the matter clear, let us state the case as plainly as possible. Human beings who clone for purposes not merely of reproduction but of selective replication are saying in effect that they are in a position to decide on the existence or non-existence of other human beings in accordance with their desires and preferences. It is one thing to say that a specific type of grape should exist and a different kind should not because of what human beings want, and quite another to say that the same should be true of a certain type of human being.

It is worth noting that this issue arises with respect both to simple cloning and to the more dramatic ambition of designer babies. The first implies that, impressed with the merits of an existing type of human being, as one might be with a particular type of sheep, we deem ourselves to be in a position to decide that there should be more of that type. The second implies that, armed with the advantages of genetic manipulation, we can rightly choose to set about creating (or more accurately 'procreating') this new type.

There are some objections that apply to the former endeavour that do not necessarily apply to the latter. If the replication of an individual is a good idea, on the grounds that more of that type is better, why not replicate several? It seems clear, for instance, that there is reason to replicate a winning racehorse or an especially productive dairy cow not just once but many times. Why not a human being, whose type is thought especially valuable? Would a dozen Mozarts not be better than two? These are intriguing questions, but I propose not to dwell on them, because more telling issues arise with respect to the basic idea of selecting 'better' human beings, whether singly or multiply.

What is the mark of 'better' here? Is it really possible to select it? And if these questions can be given clear answers, is that enough to justify us in an attempt to design babies? Take the second question first. This is in large part an empirical question, but one of considerable complexity. Could we in fact select, for the purposes of cloning, those characteristics we deem to be 'better'? Any attempt to give a positive answer to this question confronts the profound difficulties touched upon in previous chapters. If the high point of genetic engineering's ambition is designer babies, this depends crucially on the adequacy of genetic explanation. If Mozart is to be successfully replicated, whether once or many times, by means of genetic cloning, it has to be the case that the explanation of Mozart's greatness lies in his genes. The arguments of previous chapters raised serious doubts about this. While it may be true that genes had something to do with Mozart's genius (though we do not as a matter of fact know this) it seems evident that they were not the whole story. Had a phenotype identical to Mozart been born into a culture that did not have the musical history his did, there is no possibility that compositions such as *The Magic Flute* would have resulted. Perhaps the same inherent genius would have shown itself in other ways. Who knows? No one can tell, or even hazard an intelligent guess about what they might have been.

There is a version of universal Darwinism which holds that, since everything is to be explained in terms of genetic makeup, the explanation of musical brilliance (and of indefinitely many other exceptional talents and achievements) must be found in the DNA of those whose talents and achievements they are. The greatest mistake in such a contention lies in its ignoring the formative character of a learning process that integrates the biological and the social. No one can play the piano without a certain physiology – fingers of the right kind, an ability to hear, a sense of rhythm and so on – and all of these are the outcome of an evolutionary history and an expression of a distinctive genome. But to think that this biological dimension is the explanation of musical ability is to mistake a necessary for a sufficient condition. Nor is the sufficient condition a mere matter of adding on a 'cultural' component to the basic biology. In the real world of real people the two are integrated. People who play an instrument have the fingers they do partly because they play an instrument. Instruments are as they are, partly because human fingers, lips and so on, are as they are.

In short, it makes no sense to identify the biological component in musicality independently, and even less to give it some sort of priority. (It is also a mistake, of course, to suppose that at some point the cultural 'takes off' from the biological, as Dawkins's conception of 'memetics' supposes.) But even if we *could* isolate specific genes that are associated with such gifts as musicality (which, to repeat, we cannot), the arguments about genetic screening in Chapter 3 showed this to be insufficient to secure their successful replication. Were we to locate such a thing, its isolation and insertion would be a hazardous undertaking, with quite unpredictable results. The interrelations of genes are simply too complex to make plausible efforts of this kind. Besides, we know that DNA is not the only agent in biological development. It is the cell itself that determines which side of the double helix, and which chromatid, is to be translated into RNA, and thus made the programme for the cell's development. This is why there is good reason to think that cloning, even for therapeutic purposes, will always be a matter of hit and miss.

The aspiration to designer babies, then, is an impossible dream, entertained only by those who, whether in hope or in fear, are ignorant of the real state of affairs in biological understanding and contemporary biotechnology. This impossibility needs to be stated carefully. It is true that some recent research suggests that it might be possible to identify the genetic basis of psychological traits, including intelligence, and if so, it would, of course, be a relatively easy matter to select among embryos accordingly. But it is a mistake to think that these are the first steps to creating 'wonderwoman and superman' (Harris 1998: Ch. 7). Most of the important variables come into play after this stage. To suppose otherwise is to subscribe to an indefensible and absurdly ambitious genetic determinism, the view that 'stored within heredity are all the joys, sorrows, loves, hates, music, art, temples, palaces, pyramids, hovels, kings, queens, paupers, bards, prophets and philosophers' (Luther Burbank quoted in Nelkin and Lindee

1995: 19). The idea that biological inheritance is everything and that we can deliberately set about producing better and better human beings by focussing on genetic makeup is not new. It is, in fact, the animating thought of the eugenics movement which was hugely influential, and respected, in the United States in the first three decades of the twentieth century. Burbank's remark quoted above appears in a book published in 1907, and in 1934, Harry H. Cook, another prominent proponent of eugenics, wrote that 'within the nucleus of the germ cell lie the most important things in the whole world, the chromosomes, which are the determiners of character and in reality responsible for our natural individuality' (*ibid.*). Eugenics was not the preserve of the crank and crackpot; its advocates included Alexander Graham Bell, John Kellogg the industrialist, and the President of Stanford University. However its association with the Nazis led to its speedy demise after the Second World War. With the advent of modern techniques, something of the same idea, though not under the same name, has revived. But though the thought of designing and creating people in the laboratory in the manner of Frankenstein may be a more haunting one than mere selective breeding, nothing in modern genetics suggests that it has moved out of the realms of fiction.

Does this end the argument? In my view, it does put an end to the practical question, but it does not entirely conclude the moral argument. Just suppose (perhaps *per impossible*) that the explanatory and technical difficulties *could* be overcome. What then? This does not seem a wholly idle question, because we know very well that empirical and technical limitations formerly thought insurmountable have been overcome before. If this were to happen in the case of human cloning, would there be any further obstacle to the ambition of fashioning the world, even the world of human beings, in accordance with our desires and preferences? Faced with this, albeit hypothetical, question, many people still have doubts.

But granted the impracticality of the project for the foreseeable future at the very least, these can only be moral doubts. It is at this point, as it seems to me, that the idea of 'playing God' comes to the fore.

Procreative responsibility and the ethics of abortion

'Where were you when I laid the earth's foundations? Tell me, if you know and understand, who fixed its dimensions.' So says God in the Book of Job (38:4). To this the aspirants of designer babies and the like answer: we may not have been there then, but thanks to the advances of scientific understanding, we are now. This reply, as we have seen, is something of an exaggeration, possibly a very great one. Science should not be confused with Scientism. It is common to hear serious scientists remark that the more they grasp about the fundamentals of biology, physics or cosmology, the greater the sense of their own ignorance and awareness of just how little they really understand. Despite this humility on the part of the greatest scientists, there is in many quarters a widespread belief that contemporary science has put a new power in the hands of human beings. It is a power, according to this common conception, that allows the refashioning of the world at a very fundamental level and in accordance with human desires. So far I have been arguing that we do not really have such power, and if this is true the modern world of biotechnology does not actually confront those ethical dilemmas that it is often said to. Nonetheless, there is considerable interest in asking what a responsible use of that power would be if we did have it. This question returns us to a topic suspended in the previous chapter, and which I labelled 'procreative responsibility'.

In discussing the possibility of a secular version of the sanctity of human life I appealed to the principle that it is wrong for any one to assume a position from which to decide whether the life of another human being is or is not worth living. Now this seems to be a presumption that is hard to avoid when the question of designer babies is raised. To choose the characteristics of a life yet to be born presupposes that we are in a position to decide that a life – the life we 'design' – is more worth living than the life that would emerge without our designing intervention. It might be replied, of course, that the two positions are not strictly the same. No one who goes in for designer babies need hold that the lives of 'undesigned' babies are not worth living. Indeed it would be very odd for them to do so since the first such designers will themselves be leading undesigned lives. Rather, all that is claimed is that we can improve upon what otherwise might be. 'Designer babies' is a loaded expression. Suppose what we are talking about is not the creation by deliberate intention of a super-race, but a less ambitious intervention that aims to ensure that someone is born without handicaps, for instance. What is wrong with 'designing' babies that do not have learning difficulties, or have a higher IQ?

It is important to remember that this issue is now being discussed in, so to speak, the realms of fantasy. We do not know anything about the relevant connections between aspects of human personality. Any attempt to 'build' a human being with a higher IQ might trigger higher levels of mental malfunction, autism say, or other forms of emotional instability. But for the sake of argument we are leaving these hugely important doubts aside and asking 'What if?'. Once again, it seems to me, the best argumentative strategy is to consider more closely contexts with which we are familiar and look for the differences and similarities with the case of genetic design. Just as I examined the similarities and differences between human cloning and the age-old practice of taking cuttings from plants, so we may usefully look at existing practices with respect to abortion.

It is not observed as often as it should be that some of the issues surrounding the ethics of cloning human beings are similar, if not identical, to the issues that surround abortion. If we return to an idea advanced earlier in this chapter – that 'playing God' is a matter of assuming the authority to decide that the life of another is not worth living – we do not need to make any necessary reference to hi-tech methods of reproduction, since 'playing God' in this sense is exactly what happens when a foetus is aborted on the grounds that it is defective. Abortion is a vexed and emotive topic, and its introduction here may be thought to run the risk of clouding more than it illuminates. For the moment, however, I want only to explore the parallels with cloning, without taking sides in the abortion debate itself.

Abortion is a way of dealing with 'unwanted' pregnancy, but it is important to see that there is a difference between pregnancies that are 'unwanted' plain and simple, and those that are unwanted for some reason connected with the foetus itself. Some pregnancies are unwanted just in the sense that the mother (the father as well, perhaps) does not wish the child to be born. In other cases, the mother (and/or others) does not want a child to be born whose life would be marred by some defect to a degree that would make it not worth living. A familiar example is anencephaly, the condition in which the foetus has only the rudimentary elements of a brain and if brought to term is likely to die within a very few days. Where an anencephalic foetus is detectable in advance, it seems right to say that its destruction is based upon the belief that the life it would have is not worth having.

Anencephaly is, so to speak, a relatively easy case. It is also a case of the second kind. What about the first, where the foetus is unwanted in the purely subjective sense? Some people believe that this fact alone is sufficient warrant for its destruction. There is this question, though; unwanted by *whom*? In most circumstances there are many potential adoptive parents who can be said to want it, and this leaves the problem of explaining why the wishes of the natural parent(s) should be paramount in determining what happens to it. But I do not propose to examine this point further, because a more common defence of abortion in cases such as these connects the desire of the parent(s) and the life of the potential child, thereby casting the justification in the same mould as the anencephaly example.

Anencephaly is not merely a relatively easy case; it is also rare. The difficulties arise when the anticipated condition is not so evidently drastic – where the life would be longer and have sufficient experiences within it that are recognizably enjoyable or satisfying – and much less rare. Consider an example that falls somewhere between the seriously defective foetus and the merely unwanted one – Down's syndrome, mentioned briefly in an earlier chapter and in a slightly different context. A child suffering from Down's syndrome can have a reasonably long life (much longer nowadays than previously, thanks to modern treatments) and more importantly perhaps an evidently happy one. There is little or no reason to think that the life of the child with Down's syndrome is unwanted by that child. When it comes to normal children, who are merely 'unwanted', we may reasonably suppose that in many cases, perhaps most, they will suffer a measure of pain from the knowledge that their natural parents would have preferred if they had never been born. But it is hard to believe that this pain will always be so intense, and so non-compensatable by the love of (say) adoptive parents, that such a life would be intolerable. Yet this is what we must do if we are to make convincing the argument for abortion in these cases. (I leave aside here the alternative arguments couched in terms of women's rights, of which Judith Jarvis Thomson's essay 'A Defense of Abortion' is one.) We would have to claim that it is wrong to bring into existence a being who would suffer the experience of knowing that it had not been wanted because such a life would not be worth living. To state the matter at a minimum, this is far from clear.

From anencephaly through Down's syndrome to the unwanted child there is clearly a spectrum, and a spectrum in which the relative ease of the judgement is correlated with three things: the frequency of the case, our power to predict, and the existence of an authoritative first person. Anencephaly is rare, the course of the anencephalic life highly predictable, and the being whose life it is is too defective to have a view of that life. This is what makes judgement in these cases easy (or at least easier). Down's syndrome is much less rare, the course of such life much less easily predicted, and people with Down's syndrome have at least some conception of the value of their own lives (the condition varies with the degree of severity). Our knowledge of the condition gives us grounds to predict its general contours, but virtually nothing of the detailed character of the individual life. Accordingly the judgement is correspondingly chancy. Purely 'unwanted' pregnancies are common, and almost nothing can be predicted with any certainty about the life of those that are brought to term, except that a being will come into existence with a first-person perspective as fully developed as our own.

It seems natural to conclude from this that, at least as far as arguments of this kind go, moral responsibility requires greater caution the further along this spectrum we move. The strength of our medical knowledge, a shared conception of what counts as minimally valuable, and the non-existence of any first-person view keeps the judgement about an anencephalic life out of the realms of the unknown. In the case of Down's syndrome, opinions differ, though to me it seems evident that the prediction is as uncertain as the existence of a first-person point of view is certain. In the case of the merely 'unwanted' pregnancy, termination requires us to ignore our certain knowledge of the resilience of human beings and their zest for life, in order to declare (absurdly) that we can predict a life that would not be worth living, and moreover, believed to be so by the person whose life it was.

'Playing God' without God

It is important to repeat that nothing has been said in the previous section about those arguments, both for and against abortion, that invoke the concept of rights – the right to life and the right to choose. In consequence, it would be erroneous to draw any fixed conclusions about the ethics of abortion itself. The appeal to rights brings another dimension to the debate that its proper treatment needs to consider. In this context I am concerned only with *consequential* arguments about the potential life in view, because it is in arguments of this type that a parallel is to be found with the genetic design of human beings. The rights and wrongs of abortion, viewed strictly in terms of the value of the life that would or would not result, have something to tell us about the (unrealistic) ambition to produce designer babies. What has emerged from the discussion is a suitably nuanced conception of 'playing God', one that we can deploy to some effect in the present context.

'Playing God' on this interpretation implies an irresponsible transgression of boundaries that ought to constrain our procreative activity if it is to remain morally responsible. In particular, it involves going beyond the parameters set by current scientific knowledge, shared normative assessments, and regard for first-person judgements of worth. It is not difficult, in my view, to show that the ambition of fashioning 'better' human beings transgresses these same boundaries. Anyone who believes that he or she can engineer an improvement in the sorts of human being who are likely to arise from more normal processes must believe, first that they can predictably secure a certain outcome, second that this outcome is demonstrably superior, and third that their judgement of its superiority transcends or overrides the first-person judgement of the alternative, non-designed person. None of these conditions could be met. The state of our knowledge, given the complexity of human biology, cannot secure the first, and in all probability will never do so. The nature of normative judgement – weighing intelligence against strength, longevity against beauty, for instance – makes nonsense of the second, and the assumption lying behind the third implies an indefensible *hubris*. It follows that producing designer babies, even were it technologically possible (which it is not), is something we should eschew.

This gives substance to a non-theological conception of 'playing God', but it does not altogether secure the idea of procreative responsibility. To see this we need to return to the Oncomouse, the mouse whose resistence to cancer is genetically engineered away in order to aid medical research. Have scientists done anything wrong in developing the Oncomouse? In the previous chapter I rehearsed one line of argument, but suspended its consideration until now. This is the claim that such a creature is not merely odd or unnatural (in the sense of not having evolved of its own accord). Rather its essence runs counter to a fundamental principle – that the biology of a thing should be so ordered as to promote and maximize its flourishing. The unfolding of the Oncomouse's biology, by contrast, results precisely in its destruction. Cancer is the *going wrong* of cells. Accordingly the Oncomouse has been created *In order to malfunction*, and it cannot be right deliberately to engineer the natural malfunctioning of things.

I remarked earlier that, at least on first acquaintance, this seems quite a powerful argument. Its cogency rests, of course, on the crucial Aristotelian principle that the biology of a thing should be so ordered as to promote and maximize its flourishing. It further appears to identify a natural wrong that we can, so to speak, read off from the facts themselves. To hold that we can do this, of course, runs counter to a widely and deeply held view that facts and values are different spheres and that an 'ought' cannot be derived from an 'is'. In other words, the attempt to 'read off' the wrongness of the Oncomouse from its biology not only invokes a contentious Aristotelian biology, but commits what is known as the 'naturalistic fallacy'. These two points go together. If a case can be made for the Aristotelianism, the accusation about committing the naturalistic fallacy fails. However, to engage in a defence of 'natural goodness' is too large an undertaking here. (For two recent defences see Foot 2001, esp. Ch. 2, and MacIntyre 1999.) Fortunately it is not strictly necessary because our concern is not directly with the cogency of the principle, but with whether, even were we to allow that there is something objectively wrong about engineering a self-destructing creature, it can be interpreted as a secular version of 'playing God'. I think the answer is that it cannot.

The fundamental idea at work in the concept of 'playing God' is *hubris*, the overstepping of some line or other. I have been arguing that we can make (some) sense of this idea in the case of designer babies, just as we can in abortion, by appealing to the limits of human knowledge, the plurality of values, and the essentially first-person perspective on the worthwhileness of a human life. The three in combination generate an explanation of something we can call procreative irresponsibility. It is hard to see that such boundaries can apply in the case of the Oncomouse. Clearly, its creation does not exceed the state of our knowledge. Its purpose is linked to one value only – human health – and the Oncomouse does not have a first-person perspective on its own life. From a secular point of view, there is only one perspective on the value of the Oncomouse, and that is a human one, and from that perspective the usefulness of the Oncomouse for medical research is the only justification we need for its production. Of course, from a religious point of view there *is* an orientation other than the human one, namely God's, and one way of articulating the anxiety that many experience when confronted with the Oncomouse and similar genetically engineered oddities, is to say that a line has been crossed between *pro*-creation, in which human beings may legitimately engage, and creation proper, which is reserved by and for God. It is not clear, of course, whether this distinction between creation and procreation would indeed rule out the production of creatures such as the Oncomouse. But I shall not pursue the matter, chiefly because this theological way of thinking is closed to many minds in the modern Western world of genetics. The point of making reference to it is to reveal that the idea of 'playing God' without God, though it amounts to more than Glover, Harris and Peters allow, is severely limited in the moral constraints it can put upon biotechnology.

Why does this matter? It matters because even the most convinced Darwinian who greets the demise of religion with enthusiasm wants, it seems, to retain an ethical perspective that transcends and can stand in judgement on the human. In a two-part review article in *The New York Review* Frederick Crews surveys a dozen books (including Michael Behe's and Kenneth Miller's) that try either to counter some aspects of Darwinism, or to represent religion and science as two independent 'magisteria' (as in Stephen J. Gould's *Rock of Ages*, another of the books reviewed). Crews will have none of it, and at one point he especially chastises Michael Ruse for trying to give a biological interpretation to the concept of original sin. Yet, at the end of his article, human sinfulness (though not by that name) is just what he himself appears to invoke.

> Darwinism, despite its radical effect on science, has yet to temper the self-centered way in which we assess our place and actions in the world. Think of the shadows now falling

> across our planet: overpopulation, pollution, dwindling and maldistributed resources, climatic disruption, new and resurgent plagues, ethnic and religious hatred, the ravaging of forests and jungles, and the consequent loss of thousands of species a year – the greatest mass extinction, it has been said, since the age of the dinosaurs. So long as we regard ourselves as creatures apart who need only repent of our personal sins to retain heaven's blessing, we won't take the full measure of our species-wide responsibility for these calamities.
>
> An evolutionary perspective, by contrast, can trace our present woes to the dawn of agriculture a thousand years ago, when, as Niles Eldredge has observed, we became 'the first species in the entire 3.8 billion-year history of life to stop living inside local systems'. Today, when we have burst from six million to six billion exploiters of a biosphere whose resilience can no longer be assumed, the time has run out for telling ourselves that we are the darlings of a deity who placed nature there for our convenience. We are the most resourceful, but also the most dangerous and disruptive, animals in this corner of the universe. A Darwinian understanding of how we got that way could be the first step toward a wider ethics commensurate with our real transgression, not against God but against Earth itself and its myriad forms of life.
>
> (Crews 2001: 55)

Crews here ignores the Christian (and Jewish) conceptions of collective sin, and assumes just one possible theology (human dominion rather than stewardship of nature). But a far more important point is this: if the argument pursued in this chapter has been sound, Darwinism *cannot* 'temper the self-centered way in which we assess our place and actions in the world'. To ask it to do so is to seek far more than it can supply. In the end, it seems, and in so far as the fear of Frankenstein will not go away, the secular mind that utterly rejects religion must rest content with a fractured image of science.

BIBLIOGRAPHY

Aunger, Robert (ed.) (2000) *Darwinizing Culture: The Status of Memetics as a Science*, Oxford University Press

Behe, Michael J. (1998) *Darwin's Black Box*, Simon and Schuster, New York

Berry, Wendell (2000) *Life is a Miracle: An Essay Against Modern Superstition*, Counterpoint, New York

Blackburn, Simon (1996) 'I Rather Think I am a Darwinian', *Philosophy* 71 pp. 605–16

Blackmore, Susan (1999) *The Meme Machine*, Oxford University Press

Buchanan, Allen, Brock, Dan W., Daniels, Norman and Wikler, Daniel (2000) *From Chance to Choice: Genetics and Justice*, Cambridge University Press

Burley, Justine (ed.) (1999) *The Genetic Revolution and Human Rights*, Oxford University Press

Caporael, Linnda R. (2001) 'Evolutionary Psychology: Towards a Unifying Theory and a Hybrid Science', *Annual Review of Psychology* 52 pp. 607–28

Chadwick, Ruth F. (ed.) (1987) *Ethics, Reproduction and Genetic Control*, Routledge, London and New York

Chalmers, A. F. (1976) *What is This Thing Called Science?*, Open University Press, Milton Keynes

Chapman, Audrey R. (1999) *Unprecedented Choices: Religious Ethics at the Frontiers of Genetic Science*, Fortress Press, Minneapolis

Crews, Frederick (2001) 'Saving Us from Darwin', *New York Review* pp. 24–7 and 51–5

Darwin, Charles (1877) *The Descent of Man and Selection in Relation to Sex*, second edition

—— (1996) *The Origin of Species*, Oxford University Press

Dawkins, Richard (1976) *The Selfish Gene*, Oxford University Press

—— (1981) 'In Defence of Selfish Genes', *Philosophy* 56 pp. 556–73
—— (1982) *The Extended Phenotype*, W. H. Freeman, Oxford
—— (1990) *The Blind Watchmaker*, Penguin, London
—— (1996) 'Universal Darwinism' in Ruse (1996)
—— (1996a) *River out of Eden*, Orion Paperbacks, London
Dennett, Daniel C. (1996), *Darwin's Dangerous Idea: Evolution and the Meanings of Life*, Penguin Books, London
Doril, Robert (1996/7) 'Review of *Darwin's Black Box*', *American Scientist* www.amsci.org/bookshelf/Leads97/Darwin
Dose, K. (1988) 'The Origins of Life: More Questions than Answers', *Interdisciplinary Science Reviews* 13
Fodor, Jerry (1998) 'Look!' (review of E. O. Wilson, *Consilience*), *London Review of Books* vol. 20 no. 21
Foot, Philippa (2001) *Natural Goodness*, Clarendon Press, Oxford
Glover, Jonathan (1984) *What Sort of People Should There Be*? Penguin Books, London
Goodfield, June (1977) *Playing God: Genetic Engineering and the Manipulation of Life*, Harper Colophon Books, New York
Graham, Gordon (1997) *The Internet: A Philosophical Inquiry*, Routledge, London and New York
—— (2001) *Evil and Christian Ethics*, Cambridge University Press
Grant, Colin (2001) *Altruism and Christian Ethics*, Cambridge University Press
Harris, John (1998) *Clones, Genes and Immortality: Ethics and the Genetic Revolution*, Oxford University Press
Ingold, Tim (2000) *The Perception of the Environment: Essays in Livelihood, Dwelling and Skill*, Routledge, London and New York
Kitcher, Phillip (1996) *The Lives to Come: The Genetic Revolution and Human Possibilities*, Simon and Schuster, New York
MacIntyre, Alasdair (1971) *Against the Self-Images of the Age: Essays on Ideology and Philosophy*, Duckworth, London
—— (1999) *Dependent Rational Animals: Why Human Beings Need the Virtues*, Duckworth, London
Mackie, J. L. (1978) 'The Law of the Jungle: Moral Alternatives and Principles of Evolution', *Philosophy* 53 pp. 455–64
—— (1981) 'Genes and Egoism', *Philosophy* 56 pp. 553–5
Mae-Wan Ho (1999) *Genetic Engineering: Dream or Nightmare?* (revised edition), Gill and Macmillan, Dublin
Manson, Neil (1999) 'The Precautionary Principle, the Catastrophe Argument and Pascal's Wager', *Ends and Means* vol. 4 no. 1 pp. 12–16

Midgley, Mary (1979) 'Gene Juggling', *Philosophy* 54 pp. 439–58
Miller, Kenneth R. (1999) *Finding Darwin's God: A Scientist's Search for Common Ground Between God and Evolution*, Cliff Street Books/HarperCollins, New York
Morris, Henry (1974) *Scientific Creationism*, Creation-Life, San Diego
Nelkin, Dorothy and Lindee, Susan (1995) *The DNA Mystique: The Gene as a Cultural Icon*, W. H. Freeman and Company, New York
Numbers, Ronald L. (1996) 'The Creationists' in Ruse (1996)
Orr, H. Allen (1996/7) 'Darwin v. Intelligent Design (Again)', *Boston Review* www-polsci.mit.edu/bostonreview/br21.6/orr
Oswalt, W. H. (1976) *An Anthropological Analysis of Food-getting Technology*, John Wiley, New York
Peters, Ted (1997) *Playing God? Genetic Determinism and Human Freedom*, Routledge, London and New York
Pinker, Steven (1997) *How the Mind Works*, Penguin Books, London
Radcliffe Richards, Janet (2001) *Human Nature after Darwin*, Routledge, London and New York
Ridley, B. K. (2001) *On Science*, Routledge, London and New York
Ridley, Matt (1999) *Genome: The Autobiography of a Species in 23 Chapters*, Fourth Estate, London
Rolston III, Holmes (1998) 'Technology versus Nature: What is Natural?', *Ends and Means* vol. 2 no. 2
—— (1999) *Genes, Genesis and God: Values and their Origins in Natural and Human History*, Cambridge University Press
Rose, Hilary and Rose, Steven (eds) (2000) *Alas, Poor Darwin*, Jonathan Cape, London
Ruse, Michael (ed.) (1996) *But is it Science: the Philosophical Question in the Creation/Evolution Controversy*, Prometheus Books, New York
Russo, Enzo and Cove, David (1998) *Genetic Engineering: Dreams and Nightmares*, Oxford University Press
Sherlock, Richard (2001) 'Three Types of Genetic Trespassing' (unpublished conference paper)
Stove, David (1992) 'A New Religion', *Philosophy* 67 pp. 233–40
—— (1994) 'So You Think You Are a Darwinian?', *Philosophy* 69 pp. 267–77
Suzuki, David and Knudtson, Peter (1989) *Genethics: the Ethics of Engineering Life*, Unwin, London
Tenner, Edward (1997) *Why Things Bite Back: Predicting the Problems of Progress*, Fourth Estate, London
Thomson, J. J. (1971) 'A Defence of Abortion', *Philosophy and Public Affairs* 1

Turney, Jon (1998) *Frankenstein's Footsteps: Science, Genetics and Popular Culture*, Yale University Press, New Haven and London
Watts, Fraser (ed.) (2000) *Christians and Bioethics*, SPCK, London
Wilson, Edward O. (1975) *Sociobiology: the New Synthesis*, Harvard University Press, Cambridge, Mass.
—— (1992) *The Diversity of Life*, Harvard University Press, Cambridge, Mass.
—— (1995) *On Human Nature*, Penguin Books, London

INDEX